PRAISE FOR EINSTEIN'S DICE AND SCHRODINGER'S CAT:

CultureLab – Best Read of 2015

"This book can be put on the reading list of those who have enjoyed *The Theory of Everything* and want to know more."

—*Physics World*

"A highly approachable book that will appeal to readers... who are interested in physics, the history of science, and the human and political aspects of scientists and their work."

—*Library Journal*

"Halpern's book has an enormous richness of detail about both men's lives and work."

—*The Observatory*

"That's a lot to cover in a single book, and the author masters this challenge most thoroughly. While the science is covered in detail, the tone and narrative are accessible to readers with all levels of mathematical and physics proficiency. The author has served science writing well by casting light on the relationship between these two pioneers of quantum physics....Indeed, there are lessons about the often-messy process of science in this book for students, scientists, and citizens alike."

—*MAA Reviews*

"With verve, Halpern explores the fragile nature of scientific collaboration.... Halpern ably explores the clashing personalities and worldviews that had physics in churning ferment during the early part of the 20th century."

—*Kirkus Reviews*

Einstein's Dice
and Schrödinger's Cat

Einstein's Dice
and Schrödinger's Cat

How Two Great Minds Battled
Quantum Randomness to Create
a Unified Theory of Physics

Paul Halpern, PhD

BASIC BOOKS

New York

Published by Basic Books, an imprint of Perseus Books, LLC,
a division of Hachette Book Group, Inc.
First paperback edition published in 2016 by Basic Books

Books published by Basic Books are available at special discounts for bulk pur-
chases in the United States by corporations, institutions, and other organizations.
For more information, please contact the Special Markets Department at
Perseus Books, 2300 Chestnut Street, Suite 200, Philadelphia, PA 19103, or
call (800) 810-4145, ext. 5000, or e-mail special.markets@perseusbooks.com.

Designed by Pauline Brown

Library of Congress Cataloging-in-Publication Data

Halpern, Paul, 1961–

Einstein's dice and Schrödinger's cat : how two great minds battled quantum
randomness to create a unified theory of physics / Paul Halpern, PhD.

pages cm

Includes bibliographical references and index.

ISBN 978-0-465-07571-3 (hardcover) — ISBN 978-0-465-04065-0 (e-book)
1. Quantum chaos. 2. Quantum theory—Philosophy. 3. Physics—Philosophy.
4. Unified field theories. 5. Einstein, Albert, 1879–1955. 6. Schrödinger, Erwin,
1887–1961. I. Title.

QC174.17.C45H35 2015

530.13'3—dc23

ISBN 978-0-465-09683-1 (paperback)

2014041325

10 9 8 7 6 5 4 3 2 1

Dedicated to the memory of Max Dresden,
my PhD advisor, whose passion for the history of
twentieth-century physics was truly inspiring

Well who am I? (This question is meant in general, the "I" not referring just to the present writer.) The Image of God, gifted with power of thought to try and understand His world. However naive my attempt at this may be, I do have to value it higher than scrutinizing Nature for the purpose of inventing a device to . . . say, avoid splashing my spectacles in eating a grapefruit, or other very handy conveniences of life.

—Erwin Schrödinger, "The New Field Theory"

Contents

Acknowledgments

I would like to acknowledge the outstanding support of my family, friends, and colleagues in helping me see this project to completion. Thanks to the faculty and staff of the University of the Sciences, including Helen Giles-Gee, Heidi Anderson, Suzanne Murphy, Elia Eschenazi, Kevin Murphy, Brian Kirschner, and Jim Cummings, and to my colleagues in the Department of Math, Physics, and Statistics and the Department of Humanities, for supporting my research and writing. I am grateful for the camaraderie of the history of science community, including the APS Forum on the History of Physics, the Philadelphia Area Center for History of Science, and the AIP Center for History of Physics. The warm support of the Philadelphia Science Writers Association, including Greg Lester, Michal Mayer, Faye Flam, Dave Goldberg, Mark Wolverton, Brian Siano, and Neil Gussman, is most appreciated. Thanks to historians of science David C. Cassidy, Diana Buchwald, Tilman Sauer, Daniel Siegel, Catherine Westfall, Robert Crease, and Peter Pesic for useful suggestions and to Don Howard for offering helpful references. I greatly appreciate the help of Schrödinger's family, including Leonhard, Arnulf, and Ruth Braunizer, in addressing questions about his life and work. I'm grateful to musician Roland Orzabal and philosopher Hilary Putnam for kindly answering questions about their work. Thanks to science writer Michael Gross for his friendly advice on German culture and language. I appreciate the encouragement of David Zitarelli, Robert Jantzen, Linda Dalrymple Henderson, Roger Stuewer, Lisa Tenzin-Dolma, Jen Govey, Cheryl Stringall, Tony Lowe, Michael LaBossiere, Peter D. Smith, Antony Ryan, David Bood, Michael Erlich, Fred Schuepfer, Pam Quick, Carolyn Brodbeck, Marlon Fuentes, Simone Zelitch, Doug Buchholz, Linda Holtzman, Mark Singer, Jeff Shuben, Jude Kuchinsky, Kris Olson, Meg and Woody Carsky-Wilson, Carie Nguyen, Lindsey Poole, Greg Smith, Joseph Maguire, Doug DiCarlo, Patrick Pham, and Vance Lehmkuhl. I offer my sincere appreciation to Ronan and Joe Mehigan for their photographs of Schrödinger locations in Dublin. Thanks to the Princeton University Library Manuscripts Division for permission to peruse the Albert Einstein Duplicate Archives and other research materials and to the American Philosophical Society Library for access to the Archive for the History of Quantum Physics. Many thanks to Barbara Wolff and the Albert Einstein Archives in Jerusalem for reviewing my quotes from Einstein's correspondence to Schrödinger. Thanks to the Royal Irish Academy for information about their proceedings. I thank the John Simon Guggenheim Foundation for a 2002 fellowship, during which I first encountered the Einstein-Schrödinger correspondence.

Thanks to my editor, T. J. Kelleher, for his outstanding guidance and useful suggestions, and to the staff of Basic Books, including Collin Tracy, Quynh Do, Betsy DeJesu, and Sue Warga, for their help. I thank my marvelous agent, Giles Anderson, for his enthusiastic support.

Special thanks to my wife, Felicia; my sons, Eli and Aden; my parents, Bernice and Stanley Halpern; my in-laws, Arlene and Joseph Finston; Richard, Anita, Jake, Emily, Alan, Beth, Tessa, and Ken Halpern; Aaron Stanbro; Lane and Jill Hurewitz; Shara Evans; and other family members for all their love, patience, advice, and support.

Allies and Adversaries

This is the tale of two brilliant physicists, the 1947 media war that tore apart their decades-long friendship, and the fragile nature of scientific collaboration and discovery.

When they were pitted against each other, each scientist was a Nobel laureate, well into middle age, and certainly past the peak of his major work. Yet the international press largely had a different story to tell. It was a familiar narrative of a seasoned fighter still going strong versus an upstart contender hungry to seize the trophy. While Albert Einstein was extraordinarily famous, his every pronouncement covered by the media, relatively few readers were conversant with the work of Austrian physicist Erwin Schrödinger.

Those following Einstein's career knew that he had been working for decades on a unified field theory. He hoped to extend the work of nineteenth-century British physicist James Clerk Maxwell in uniting the forces of nature through a simple set of equations. Maxwell had provided a unified explanation for electricity and magnetism, called electromagnetic fields, and identified them as light waves. Einstein's own general theory of relativity described gravity as a warping of the geometry of space and time. Confirmation of the theory had won him fame. However, he didn't want to stop there. His dream was to incorporate Maxwell's results into an extended form of general relativity and thereby unite electromagnetism with gravity.

Every few years, Einstein had announced a unified theory to great fanfare, only to have it quietly fail and be replaced by another. Starting in the late 1920s, one of his primary goals was a deterministic alternative to probabilistic quantum theory, as developed by Niels Bohr, Werner Heisenberg, Max Born, and others. Although he realized that quantum theory was experimentally successful, he judged it incomplete.

In his heart he felt that "God did not play dice," as he put it, couching the issue in terms of what an ideal mechanistic creation would be like. By "God" he meant the deity described by seventeenth-century Dutch philosopher Baruch Spinoza: an emblem of the best possible natural order. Spinoza had argued that God, synonymous with nature, was immutable and eternal, leaving no room for chance. Agreeing with Spinoza, Einstein sought the invariant rules governing nature's mechanisms. He was absolutely determined to prove that the world was absolutely determined.

Exiled in Ireland in the 1940s after the Nazi annexation of Austria, Schrödinger shared Einstein's disdain for the orthodox interpretation of quantum mechanics and saw him as a natural collaborator. Einstein similarly found in Schrödinger a kindred spirit. After sharing ideas for unification of the forces, Schrödinger suddenly announced success, generating a storm of attention and opening a rift between the men.

You may have heard of Schrödinger's cat—the feline thought experiment for which the general public knows him best. But back when this feud took place, few people outside of the physics community had heard of the cat conundrum or of him. As depicted in the press, he was just an ambitious scientist residing in Dublin who might have landed a knockout punch on the great one.

The leading announcer was the *Irish Press,* from which the international community learned about Schrödinger's challenge. Schrödinger had sent them an extensive press release describing his new "theory of everything," immodestly placing his own work in the context of the achievements of the Greek sage Democritus (the coiner of the term "atom"), the Roman poet Lucretius, the French philosopher Descartes, Spinoza, and Einstein himself. "It is not a very becoming thing for a scientist to advertise his own discoveries," Schrödinger told them. "But since the Press wishes it, I submit to them."[1]

The *New York Times* cast the announcement as a battle between a maverick's mysterious methods and the establishment's lack of progress. "How Schrödinger has proceeded we are not told," it reported.[2]

For a fleeting moment it seemed that a Viennese physicist whose name was then little known to the general public had beaten the great Einstein to a theory that explained everything in the universe. Perhaps it was time, puzzled readers may have thought, to get to know Schrödinger better.

A Gruesome Conundrum

Today, what comes to mind for most people who have heard of Schrödinger are a cat, a box, and a paradox. His famous thought experiment, published as part of a 1935 paper, "The Present Situation in Quantum Mechanics," is one of the most gruesome devised in the history of science. Hearing about it for the first time is bound to trigger gasps of horror, followed by relief that it is just a hypothetical experiment that presumably has never been attempted on an actual feline subject.

Schrödinger proposed the thought experiment in 1935 as part of a paper that investigated the ramifications of entanglement in quantum physics. Entanglement (the term was coined by Schrödinger) is when the condition of two or more particles is represented by a single quantum state, such that if something happens to one particle the others are instantly affected.

Inspired in part by dialogue with Einstein, the conundrum of Schrödinger's cat presses the implications of quantum physics to their very limits by asking us to imagine the fate of a cat becoming entangled with the state of a particle. The cat is placed in a box that contains a radioactive substance, a Geiger counter, and a sealed vial of poison. The box is closed, and a timer is set to precisely the interval at which the substance would have a 50–50 chance of decaying by releasing a particle. The researcher has rigged the apparatus so that if the Geiger counter registers the click of a single decay particle, the vial would be smashed, the poison released, and the cat dispatched. However, if no decay occurs, the cat would be spared.

According to quantum measurement theory, as Schrödinger pointed out, the state of the cat (dead or alive) would be entangled with the state of the Geiger counter's reading (decay or no decay) until the box is opened. Therefore, the cat would be in a zombielike quantum superposition of deceased and living until the timer went off, the researcher opened the box, and the quantum state of the cat and counter "collapsed" (distilled itself) into one of the two possibilities.

From the late 1930s until the early 1960s the thought experiment was little mentioned, except sometimes as a classroom anecdote. For instance, Columbia University professor and Nobel laureate T. D. Lee would tell the tale to his students to illustrate the strange nature of quantum collapse.[3] In 1963, Princeton physicist Eugene Wigner mentioned the thought experiment in a piece he wrote about quantum measurement and extended it into what is now referred to as the "Wigner's friend" paradox.

Renowned Harvard philosopher Hilary Putnam—who learned about the conundrum from physicist colleagues—was one of the first scholars outside of the world of physics to analyze and discuss Schrödinger's thought experiment.[4] He described its implications in his classic 1965 paper "A Philosopher Looks at Quantum Mechanics," published as a book chapter. When the paper was mentioned the same year in a *Scientific American* book review, the term "Schrödinger's cat" entered the realm of popular science. Over the decades that followed, it crept into culture as a symbol of ambiguity and has been mentioned in stories, essays, and verse.

Despite the public's current familiarity with the cat paradox, the physicist who developed it still isn't well known otherwise. While Einstein has been an icon since the 1920s, the very emblem of a brilliant scientist, Schrödinger's life story is scarcely familiar. That is ironic because the adjective "Schrödinger's"—in the sense of a muddled existence—could well have applied to him.

A Man of Many Contradictions

The ambiguity of Schrödinger's cat perfectly matched the contradictory life of its creator. The bookish, bespectacled professor maintained a quantum superposition of contrasting views. His yin-yang existence began in his youth when he learned German and English from different family members and was raised bilingual. With ties to many countries but a supreme love of his native Austria, he never felt comfortable with either nationalism or internationalism and preferred avoiding politics altogether.

An enthusiast of fresh air and exercise, he would drown others in the smoke from his omnipresent pipe. At formal conferences, he'd walk in dressed like a backpacker. He'd call himself an atheist and talk about divine motivations. At one point in his life he lived with both his wife and another woman who was the mother of his first child. His doctoral work was a mixture of experimental and theoretical physics. During the early part of his career he briefly considered switching to philosophy before veering back to science. Then came whirlwind shifts between numerous institutions in Austria, Germany, and Switzerland.

As physicist Walter Thirring, who once worked with him, described, "It was like he was always being chased: from one problem to another by his genius, from one country to another by the political powers in the twentieth century. He was a man full of contradictions."[5]

At one point in his career, he argued vehemently that causality should be rejected in favor of pure chance. Several years later, after developing the deterministic Schrödinger equation, he had second thoughts. Perhaps there are causal laws after all, he argued. Then physicist Max Born reinterpreted his equation probabilistically. After fighting that reinterpretation, he started to sway back toward the chance conception. Later in life, his philosophical roulette wheel landed once again in the direction of causality.

In 1933, Schrödinger heroically gave up an esteemed position in Berlin because of the Nazis. He was the most prominent non-Jewish physicist to leave of his own accord. After working in Oxford, he decided to move back to Austria and became a professor at the University of Graz. But then, strangely enough, after Nazi Germany annexed Austria, he tried to cut a deal with the government to keep his job. In a published confession, he apologized for his earlier opposition and proclaimed his loyalty to the conquering power. Despite his pandering, he had to leave Austria anyway, moving on to a prominent position at the newly founded Dublin Institute for Advanced Studies. Once on neutral ground, he recanted his self-renunciation.

"He demonstrated impressive civil courage after Hitler came to power in Germany and . . . left the most prominent German professorship in physics," noted Thirring. "As the Nazis caught up with him, he was forced into a pathetic show of solidarity with the terror regime."[6]

Quantum Comrades

Einstein, who had been a colleague and dear friend in Berlin, stuck by Schrödinger all along and was delighted to correspond with him about their mutual interests in physics and philosophy. Together they battled a common villain: sheer randomness, the opposite of natural order.

Schooled in the writings of Spinoza, Schopenhauer—for whom the unifying principle was the force of will, connecting all things in nature—and other philosophers, Einstein and Schrödinger shared a dislike for including ambiguities and subjectivity in any fundamental description of the universe. While each played a seminal role in the development of quantum mechanics, both were convinced that the theory was incomplete. Though recognizing the theory's experimental successes, they believed that further theoretical work would reveal a timeless, objective reality.

Their alliance was cemented by Born's reinterpretation of Schrödinger's wave equation. As originally construed, the Schrödinger equation was designed to model the continuous behavior of tangible matter waves, representing electrons in and out of atoms. Much as Maxwell constructed deterministic equations describing light as electromagnetic waves traveling through space, Schrödinger wanted to create an equation that would detail the steady flow of matter waves. He thereby hoped to offer a comprehensive accounting of all of the physical properties of electrons.

Born shattered the exactitude of Schrödinger's description, replacing matter waves with probability waves. Instead of physical properties being assessed directly, they needed to be calculated through mathematical manipulations of the probability waves' values. In doing so, he brought the Schrödinger equation in line with Heisenberg's ideas about indeterminacy. In Heisenberg's view, certain pairs of physical quantities, such as position and momentum (mass times velocity) could not be measured simultaneously with high precision. He encoded such quantum fuzziness in his famous uncertainty principle: the more precisely a researcher measures a particle's position, the less precisely he or she can know its momentum—and the converse.

Aspiring to model the actual substance of electrons and other particles, not just their likelihoods, Schrödinger criticized the intangible elements of the Heisenberg-Born approach. He similarly eschewed Bohr's quantum philosophy, called "complementarity," in which either wavelike or particlelike properties reared their heads, depending on the experimenter's choice of measuring apparatus. Nature should be visualizable, he rebutted, not an inscrutable black box with hidden workings.

As Born's, Heisenberg's, and Bohr's ideas became widely accepted among the physics community, melded into what became known as the "Copenhagen interpretation" or orthodox quantum view, Einstein and Schrödinger became natural allies. In their later years, each hoped to find a unified field theory that would fill in the gaps of quantum physics

Portrait of Albert Einstein in his later years.
Courtesy of the University of New Hampshire, Lotte Jacobi Collection, and the AIP Emilio Segre Visual Archives, donated by Gerald Holton.

and unite the forces of nature. By extending general relativity to include all of the natural forces, such a theory would replace matter with pure geometry—fulfilling the dream of the Pythagoreans, who believed that "all is number."

Schrödinger had good reason to be much indebted to Einstein. A talk by Einstein in 1913 help spark his interest in pursuing fundamental questions in physics. An article Einstein published in 1925 referenced French physicist Louis de Broglie's concept of matter waves, inspiring Schrödinger to develop his equation governing the behavior of such waves. That equation earned Schrödinger the Nobel Prize, for which Einstein, among others, had nominated him. Einstein endorsed his appointment as a professor at the University of Berlin and as a member of the illustrious Prussian Academy of Sciences. Einstein warmly invited Schrödinger to his summer home in Caputh and continued to offer guidance in their extensive correspondence. The EPR thought experiment, developed by Einstein and his assistants Boris Podolsky and Nathan Rosen to illustrate murky aspects of quantum entanglement, along with a suggestion by Einstein about a quantum paradox involving gunpowder, helped inspire Schrödinger's cat conundrum. Finally, the ideas developed by Schrödinger in his quest for unification were variations of proposals by Einstein. The two theorists frequently corresponded about ways to tweak general relativity to make it mathematically flexible enough to encompass other forces besides gravity.

Portrait of a Fiasco

Dublin's Institute for Advanced Studies, where Schrödinger was the leading physicist throughout the 1940s and early 1950s, was modeled directly on Princeton's Institute for Advanced Study, where Einstein had played the same role since the mid-1930s. Irish press reports often compared the two of them, treating Schrödinger as Einstein's Emerald Isle equivalent.

Schrödinger took every opportunity to mention his connection with Einstein, going so far as to reveal some of the contents of their private

Erwin Schrödinger, in midlife, relaxing outdoors. *Photo by Wolfgang Pfaundler, Innsbruck, Austria, courtesy AIP Emilio Segre Visual Archives*

correspondence when it suited his purpose. For example, in 1943, after Einstein wrote personally to Schrödinger that a certain model for unification had been the "tomb of his hopes" in the 1920s, Schrödinger exploited that statement to make it look like he had succeeded where Einstein had failed. He read the letter publicly to the Royal Irish Academy, bragging that he had "exhumed" Einstein's hopes through his own calculations. The lecture was reported in the *Irish Times,* capped by the misleading headline "Einstein Tribute to Schroedinger."[7]

At first Einstein graciously chose to ignore Schrödinger's boasts. However, the press reaction to a speech Schrödinger gave in January 1947 claiming victory in the battle for a theory of everything proved too much. Schrödinger's bold statement to the press asserting that he had achieved the goal that had eluded Einstein for decades (by developing a theory that superseded general relativity) was flung in Einstein's face, in hopes of spurring a reaction.

And react he did. Einstein's snarky reply reflected his deep displeasure with Schrödinger's overreaching assertions. In his own press release, translated into English by his assistant Ernst Straus, he responded: "Professor Schroedinger's latest attempt . . . can . . . be judged only on the basis of its mathematical qualities, but not from the point of view of 'truth' and agreement with facts of experience. Even from this point of view, it can see no special advantages—rather the opposite."[8]

The bickering was reported in newspapers such as the *Irish Press,* which conveyed Einstein's admonition that it is "undesirable . . . to present such preliminary attempts to the public in any form. It is even worse when the impression is created that one is dealing with definite discoveries concerning physical reality."[9]

Humorist Brian O'Nolan, writing in the *Irish Times* under the nom de plume "Myles na gCopaleen," savaged Einstein's response, in essence calling him arrogant and out of touch. "What does Einstein know of the use and meaning of words?" he wrote. "Very little, I should say. . . . For instance what does he mean by terms like 'truth' and 'the facts of experience.' His attempt to meet shrewd newspaper readers on their own ground is not impressive."[10]

These two old friends, comrades in the battle against the orthodox interpretation of quantum mechanics, had never anticipated that they would be battling in the international press. That was certainly neither Schrödinger nor Einstein's intention when they had begun their correspondence about unified field theory some years earlier. However,

Schrödinger's audacious claims to the Royal Irish Academy proved irresistible to eager reporters, who often trawled for stories related to Einstein.

One impetus for the skirmish was Schrödinger's need to please his host, Irish *taoiseach* (prime minister) Éamon de Valera, who had personally arranged for his journey to Dublin and appointment to the Institute. De Valera took an active interest in Schrödinger's accomplishments, hoping that he would bring glory to the newly independent Irish republic. As a former math instructor, de Valera was an aficionado of Irish mathematician William Rowan Hamilton. In 1943, he made sure that the centenary of one of Hamilton's discoveries, a type of numbers called quaternions, was honored throughout Ireland. Much of Schrödinger's work made use of Hamilton's methods. What better way to honor liberated Ireland and its leading light, Hamilton, by bringing it newfound fame as the place where Einstein's relativity was dethroned and replaced with a more comprehensive theory? Schrödinger's far-reaching pronouncement matched his patron's hopes perfectly. The *Irish Press,* owned and controlled by de Valera, made sure the world knew that the land of Hamilton, Yeats, Joyce, and Shaw could also produce a "theory of everything."

Schrödinger's approach to science (and indeed to life) was impulsive. Feeling blessed with promising results, he wanted to trumpet them to the world, not realizing until it was too late that he was slighting one of his dearest friends and mentors. He considered his discovery—purportedly a simple mathematical way of encapsulating the entirety of natural law—to be something like a divine revelation. Therefore, he was anxious to divulge what he saw as a fundamental truth revealed only to him.

Needless to say, Schrödinger came nowhere near developing a theory that explained everything, as Einstein correctly pointed out. He merely found one of many mathematical variations of general relativity that technically made room for other forces. However, until solutions to that variation could be found that matched physical reality, it was just an abstract exercise rather than a genuine description of nature. While there are myriad ways to extend general relativity, none has been found so far that matches how elementary particles actually behave, including their quantum properties.

In the hype department, though, Einstein was hardly an innocent bystander. Periodically he had proposed his own unification models and overstated their importance to the press. For example, in 1929, he announced to great fanfare that he had found a theory that united the forces of nature and surpassed general relativity. Given that he hadn't found (and wouldn't find) physically realistic solutions to his equations, his announcement was extremely premature. Yet he criticized Schrödinger for essentially doing the same thing.

Schrödinger's wife, Anny, later revealed to physicist Peter Freund that he and Einstein were each contemplating suing the other for plagiarism. Physicist Wolfgang Pauli, who knew both of them well, warned them of the possible consequences of pursuing legal remedies. A lawsuit played out in the press would be embarrassing, he advised them. It would quickly degenerate into a farce, sullying their reputations. Their acrimony was such that Schrödinger once told physicist John Moffat, who was visiting Dublin, "my method is far superior to Albert's! Let me explain to you, Moffat, that Albert is an old fool."[11]

Freund speculated about the reasons two aging physicists would seek a theory of everything. "One can answer this question on two levels," he said. "On one level it is an act of ultimate grandiosity. . . . [They] were extremely successful in physics. As they see their powers waning, they take one final stab at the biggest problem: finding the ultimate theory, *ending* physics. . . . On another level, maybe these men are just driven by the same insatiable curiosity that has stood them in such good stead in their youth. They want to know the solution to the puzzle that has preoccupied them throughout life; they want to have a glimpse of the promised land in their lifetime."[12]

Frayed Unity

Many physicists spend their careers focused on very specific questions about particular aspects of the natural world. They see the trees, not the forest. Einstein and Schrödinger shared much broader aspirations. Through their readings of philosophy, each was convinced that nature had a grand blueprint. Their youthful journeys led them to significant discoveries—including Einstein's theory of relativity and Schrödinger's wave equation—that revealed part of the answer. Tantalized by part of the solution, they hoped to complete their life missions by finding a theory that explained everything.

However, as in the case of religious sectarianism, even minor differences in outlook can lead to major conflicts. Schrödinger jumped the gun because he thought he had miraculously found a clue that Einstein somehow had missed. His false epiphany, together with the performance pressures he faced because of his academic position, generated an impulsive need to come forward before he had gathered enough proof to confirm his theory.

Their skirmish came at a cost. From that point on, their dream of cosmic unity was tainted with personal conflict. They squandered the prospect of spending their remaining years in friendly dialogue, headily discussing possible clockwork mechanisms of the universe. Having waited billions of years for a complete explanation of its workings, the cosmos would be patient, but two great thinkers had lost their fleeting opportunity.

CHAPTER ONE

The Clockwork Universe

These transient facts,
These fugitive impressions.
Must be transformed by mental acts,
To permanent possessions.
Then summon up your grasp of mind,
Your fancy scientific,
Till sights and sounds with thought combined
Become of truth prolific.

—JAMES CLERK MAXWELL, from "To the
Chief Musician upon Nabla: A Tyndallic Ode"

Until the age of relativity and quantum mechanics, the two greatest unifiers of physics were Isaac Newton and James Clerk Maxwell. Newton's laws of mechanics demonstrated how the changing motions of objects were governed by their interactions with other objects. His law of gravitation codified one such interaction; the force causing celestial bodies, such as the planets, to follow particular paths, such as elliptical orbits. He brilliantly showed how all manner of phenomena on Earth—an arrow's trajectory, for instance—find explanation in a universal picture.

Newtonian physics is completely deterministic. If, at a particular instant, you knew the positions and velocities of every object in the universe, along with all the forces on them, you could theoretically predict their complete behavior forever. Inspired by the power of Newton's laws, many nineteenth-century thinkers believed that only practical limitations, such as the daunting challenge of gathering colossal amounts of data, prevented scientists from perfectly prognosticating everything.

Randomness, from that strictly deterministic perspective, is an artifact of complex situations involving a large number of components and a medley of different environmental factors. Take, for example, the quintessential "random" act of tossing a coin. If a scientist could painstakingly map out all the air currents affecting the coin and knew the precise speed and angle of its launch, in principle he or she would be able to predict its spin and trajectory. Some staunch determinists would go so far as to say that if enough information were known about the person's background and prior experiences, the thoughts of the individual tossing the coin could be predicted as well. In that case a researcher could anticipate the brain patterns, nerve signals, and muscle contractions triggering the toss, making its outcome even more predictable. In short, believers in the standpoint that the entire universe runs like a perfect clock dismiss the notion that anything is fundamentally random.

Indeed, on astronomical scales, such as the domain of the solar system, Newton's laws are remarkably accurate. They wonderfully reproduce German astronomer Johannes Kepler's laws describing how planets orbit the Sun. Our capacity to anticipate celestial events, such as solar eclipses and planetary conjunctions, and to launch spacecraft precisely toward faraway targets are testimony to the clockwork predictability of Newtonian mechanics, particularly as applied to gravitation.

Maxwell's equations brought unity to another natural force, electromagnetism. Before the nineteenth century, science treated electricity and magnetism as separate phenomena. However, experimental work by British physicist Michael Faraday and others demonstrated a deep connection, and through simple mathematical relationships Maxwell cemented the link. His four equations show precisely how the changing motion of electric charges and currents leads to energetic oscillations that radiate through space as electromagnetic waves. The relationships are the epitome of mathematical conciseness, compact enough to fit on a T-shirt yet powerful enough to describe all manner of electromagnetic phenomena. Through his pairing of electricity and magnetism, Maxwell pioneered the notion of unification of the forces.

Today we know that the four fundamental forces of nature are gravitation, electromagnetism, and the strong and weak nuclear interactions. We believe that all other forces (friction, for instance) are derived from that quartet. Each of the four operates at a different scale and possesses a different strength. Gravitation, the weakest force, draws massive bodies together over wide distances. Electromagnetism is far, far stronger and affects charged objects. Although it operates at similarly long range, its effect is reduced by the fact that almost everything in space is electrically neutral. The strong interaction operates on the nuclear scale, binding together certain types of subatomic particles (those built from quarks, such as protons and neutrons). The weak interaction, operating in the same realm, affects nuclei, causing certain types of radioactive decay. Maxwell's fusion inspired subsequent thinkers, such as Einstein and Schrödinger, in their attempts to achieve even greater unification.

Unlike conventional wave types, as Maxwell demonstrated, electromagnetic waves can propagate without a material medium. In 1865, he calculated the speed by which electromagnetic waves travel through the vacuum of space and found it to be identical to that of light. He thereby concluded that electromagnetic and light waves (including invisible forms of light such as radio waves) are one and the same.

Like Newtonian physics, Maxwellian physics is wholly deterministic: jiggle a charge in a transmitting antenna and you can predict the signal picked up by a receiving antenna. Radio stations depend on such reliability.

Unfortunately, Maxwell's unity did not quite match up with Newton's unity. The two theories offered clashing predictions for how the speed of light would appear to a moving observer. While Maxwell's equations mandated its constancy, Newton's laws predicted that its relative speed would depend upon the observer's speed. Yet both answers seemed reasonable. Coincidentally, the solver of the riddle would be born in the year of Maxwell's death.

The Compass and the Dance

In Ulm, Germany, on March 14, 1879, Pauline Einstein (née Koch), the wife of Hermann Einstein, an electrical engineer, gave birth to their first child, Albert. The young boy spent little time in that quaint Swabian city. As one of many affected by Maxwell's revolution, Hermann soon brought the family to bustling Munich, where he cofounded an electrification business. There Albert's sister, Maja, was born.

Albert's exposure to the notion of magnetic attraction came early in life. At the age of five, he was sick in bed one day when his father gave him a compass as a present. Turning the shiny instrument in his hand, the young boy marveled at its wondrous properties. Somehow its needle mysteriously knew the way back to its starting place, marked "N." His mind raced to find a missing cause for such odd behavior.

While Einstein never had a younger brother, he would someday refer to a kindred Austrian as the closest equivalent. On August 12, 1887, in the Vienna district of Erdberg, Erwin Schrödinger was born. He was the only child of Rudolf Schrödinger, who had originally studied chemistry, and Georgine "Georgie" Bauer Schrödinger, the Anglo-Austrian daughter of accomplished chemist Alexander Bauer (Rudolf's professor).

Rudolf had inherited a lucrative business manufacturing linoleum and oilcloths. His true passion, though, was in science and the arts, especially botany and painting. He bestowed on Erwin a sense that an educated man should have diverse pursuits and a love of culture.

Young Erwin was very close to his mother's younger sister Minnie. From very early on, Aunt Minnie was his confidante and advisor about worldly subjects. He was curious about everything, and even before he could read or write, he dictated his impressions to her, and she loyally jotted them down.

According to Minnie's recollections, Erwin was particularly fond of astronomy. When he was around four years old, he loved playing a game that illustrated planetary motion. Little Erwin would run around Aunt Minnie in circles, acting like he was the Moon and she was Earth. Then they would walk slowly around a lamp, pretending that it was the Sun. Looping around his aunt as they orbited the glowing fixture, he would experience the intricacy of lunar motion firsthand.

Einstein's childhood fascination with a compass and Schrödinger's "dance of the planets" foreshadowed their later interests in electromagnetism and gravitation, the two fundamental forces recognized at that time. The youths shared the prevailing belief that nature seemed clockwork in its precise mechanisms. Later in life they would strive to find a greater unity that included both forces and was similarly mechanistic.

Each would began their careers along practical lines, following their fathers in looking at applications of science to everyday life, but veered toward loftier aspirations as their lives progressed. In time, each became obsessed with unraveling the mysteries of the universe, trying to discern its fundamental principles. Each was extraordinarily gifted in the insights and calculations needed for theoretical physics.

Each hoped to follow in the footsteps of Newton and Maxwell in formulating new equations describing the natural world. Indeed, some of the most important equations of twentieth-century physics would be developed by and named after the two men. In assessing hypotheses, particularly during the late stages of their careers, each would rely heavily on philosophical considerations, drawing upon thinkers such as Spinoza, Schopenhauer, and Ernst Mach. Inspired by Spinoza's concept of God as an immutable natural order, they sought a simple, invariant set of rules governing reality. Intrigued by Schopenhauer's notion that the world is shaped by a single, driving principle called "will," they looked for grand unifying schemes. Motivated by Mach's idea that science should be tangible, they eschewed hidden processes, such as unseen, nonlocal quantum connections, in favor of manifest, causal mechanisms.

To spend days, months, or years obsessed with finding the simplest mathematical formulas that comprehensively describe certain aspects of nature requires something like a religious fervor. The ultimate equations were their holy grail, their Kabbalah, and their philosopher's stone. Judgments about what makes an equation elegant and beautiful often stem from a deep-seating sense of cosmic order. While neither Einstein or Schrödinger was religious in the traditional sense—Einstein was Jewish and Schrödinger was of Lutheran and Catholic heritage, but neither professed faith or attended religious services—they shared a wonder about the organizing principles for the universe and how these are expressed mathematically. Each had a passion for mathematics, not for its own sake, but as a tool for understanding nature's guiding laws.

From where does a lifelong interest in mathematics arise? Sometimes it is as simple as the elegant diagrams and logical proofs set out in a geometry primer.

Strange Parallels

In 1891, when Einstein was twelve and attending Luitpold Gymnasium (secondary school), he acquired a geometry book. In his mind it was a wonder comparable to the compass—introducing him to a comforting kind of order that transcended the jumble of everyday experience. Hardly just a text, for him it was a "holy book," as he later described it. Proofs based upon firm, undisputable statements showed that despite the clatter of horse-drawn streetcars, the shamble of sausage vending carts, and the din of festive beer drinkers in Munich, underlying the world was a quiet, unwavering truth. "This lucidity and certainty made an indescribable impression upon me," he recalled.[1]

Some of the assertions made in the book seemed obvious to him. He had earlier learned about the Pythagorean theorem for right-angled triangles: the sum of the squares of the two perpendicular sides was equal to the square of the third side, the hypotenuse. The book showed how if you vary one of the acute (smaller than 90-degree) angles, the lengths of the sides must change too. That seemed clear to him, even without proof.

However, other geometric propositions were not self-evident. Einstein welcomed the primer's methodical treatment of theorems that didn't seem obvious but turned out to be true—such as that the altitudes of a triangle (perpendicular line segments from each side to a corner) must meet at a point. He didn't mind that the proofs in the book were ultimately based upon unproven statements called axioms (common notions) and postulates (notions specific to a particular field). He was eager to pay the price of unquestioned acceptance of a handful of axioms for a bounty of proven conjectures.

The plane geometry described in the book could be traced back more than two thousand years to the work of the Greek mathematician Euclid. Euclid's *Elements* organized geometric knowledge into dozens of proven theorems and corollaries. These are derived systematically from a set of five axioms and five postulates. While each of the axioms and postulates was meant to be a self-evident truth, such as a part being smaller than a whole and that if two things are equal to a third thing they are equal to each other, the fifth postulate, relating to angles, doesn't seem quite so obvious.

"If two . . . lines meet a third line, so as to make the sum of the two interior angles on the same side less than two right angles, these lines being produced shall meet at some finite distance."[2] In other words, draw three lines such that the first two intersect the third at angles facing each other that are less than 90 degrees each. Eventually, if you extend them far enough, the first two lines must intersect and form a triangle. So, for instance, if one angle is 89 degrees and the other facing angle is 89 degrees, there must be a third angle (of 2 degrees) where the first two lines meet—making a very stretched-out triangle.

Mathematicians speculate that the fifth postulate was placed last on the list because Euclid tried to prove it from the other axioms and postulates but couldn't. Indeed, Euclid managed to generate fully twenty-eight theorems using the four other postulates before he added the fifth into the mix. It was like an expert keyboardist banging out all the music for twenty-eight songs at a concert before finding the need to borrow an acoustic guitar to create just the right sound for the twenty-ninth. Sometimes the instruments at hand aren't enough to complete a piece and one must improvise by bringing in another.

Euclid's fifth postulate has come to be known as the "parallel postulate" mainly because of the work of Scottish mathematician John Playfair. Playfair developed another version of the fifth postulate that, while not completely logically equivalent to the original, serves a similar purpose in proving theorems. In Playfair's version, for every line and a point not on it, there is exactly one line through the point that is parallel to the original line.

Over the centuries, various attempts have been made to prove the fifth postulate—either in Euclid's or Playfair's rendition—from the other postulates. Even the famed Persian poet and philosopher Omar Khayyam tried to no avail to transform that postulate into a proved theorem. Eventually the mathematical community concluded that the postulate was wholly independent and gave up the idea of proving it.

When young Einstein perused the geometry book, he was unaware of the controversies surrounding the parallel postulate. Furthermore, he shared the centuries-old idea that Euclidean geometry was sacrosanct. The laws and proofs seemed as solid, timeless, and majestic as the Bavarian Alps.

However, far north of Munich, in the quaint university town of Göttingen, mathematicians were engaging in a bold experiment to remake the field of geometry. The cobblestoned sanctuary for cerebral life had become an enclave for a radical rethinking of mathematics called non-Euclidean geometry. The novel geometric approach bore as much resemblance to the traditional variety as psychedelic Peter Max posters do to Rembrandt's work. As Einstein was learning the old-school rules for points, lines, and shapes on flat planes, brilliant mathematicians such as Felix Klein—recruited to Göttingen from Leipzig—were promoting a far more flexible playbook involving relationships within curved and twisted surfaces. Klein's most mind-blowing creation, the Klein bottle, is something like a vase in which the inside and outside surfaces are connected via a twist in a higher dimension. Such a monstrosity would not be found yet in primers, where Euclid's ironclad rules locked such horrors out. Yet Klein showed that Euclidean and non-Euclidean geometries are equally valid. By the 1890s, his revolutionary vision helped open up the once staid geometry club to freaks as well as squares.

Non-Euclidean geometry is not just a free-for-all, however. Like its predecessor, it has its own regulations. The essence of non-Euclidean geometry is to replace the parallel postulate with novel assertions while keeping all the other postulates the same. It recognizes that since the parallel postulate is independent, it is in some way dispensable, opening up the door to radical new options.

Mathematician Carl Friedrich Gauss was the first to propose a non-Euclidean geometry, although he did not publish those initial thoughts. In Gauss's version, later dubbed by Klein "hyperbolic geometry," the parallel postulate is replaced with the idea that through any point not on a line there are an infinite number of lines through that point parallel to the original line. One can think of it as something like clenching the end of a paper fan tightly just above a long, narrow table. If the table represents a line and your hand a point not on the line, then the folds of the fan demonstrate the myriad lines through the line that do not intersect the original line. The term "hyperbolic" derives from the shape of the fanning out of parallel lines being akin to the branches of a hyperbola.

Gauss noted a curious thing about triangles situated in a hyperbolic geometry: the sum of their angles is less than 180 degrees. In contrast, the angles of Euclidean triangles inevitably add up to precisely 180 degrees, such as an isosceles right triangle with two 45-degree angles and one 90-degree angle. The imaginative artist M. C. Escher would later tap into this distinction to create curious patterns of distorted sub-180-degree triangles living in a hyperbolic reality.

One way of picturing hyperbolic geometry is to imagine points, lines, and shapes etched on a saddle-shaped surface instead of a flat plane. If your tastes are more epicurean than equestrian, a curvy potato chip would do just fine. The saddle shape naturally causes nearby lines to veer away from each other. As much as they would "like" to be straight, sets of parallel lines bend away from each other, making it easier for them to avoid each other. This permits an unlimited number of lines through each point to be parallel to lines not going through that point. Moreover, the saddle-shape squeezes the corners of triangles, rendering the sum of their angles less than 180 degrees.

In another variation of non-Euclidean geometry, first proposed by Gauss's student Bernhard Riemann in 1854, published in 1867, and later designated by Klein's term "elliptic geometry," the parallel postulate is replaced with a rule that eliminates the possibility of parallel lines altogether. For every point outside a line, it states, there are no lines through that point parallel to the original line. In other words, all the lines through that point must intersect the original line somewhere in space. Riemann showed that lines on spherical surfaces possess that property.

If the idea of no parallel lines seems strange, think of Earth. Each of its lines of longitude intersects all the others at the North and South Poles. Thus if one ambitious traveler starts in downtown Toronto, journeys northward along its main thoroughfare, Yonge Street, hires a dogsled and icebreaking boat, and keeps going until she reaches the North Pole, while her sister takes a similar route starting from Moscow, their paths would seem parallel at first, but the siblings would inevitably meet up.

Curiously, such a ban on parallels serves to transform the nature of triangles in yet another way. In elliptic geometry, the sum of a triangle's angles add up to more than 180 degrees. Indeed, a triangle can be formed with all right angles, making its angular sum a full 270 degrees. For example, the triangle composed of the 0-degree and 90-degree lines of longitude, along with the part of the equator that connects them, has three perpendicular sides.

Riemann developed very sophisticated mathematical machinery to analyze curved surfaces in any number of dimensions; these surfaces came to be called manifolds. Riemann showed how the differences between curved and flat spaces could be pinned down from point to point using what is now called the Riemann curvature tensor. A tensor is a mathematical entity that alters in a particular way during coordinate transformations. He showed that there are three main ways space can be curved—positive curvature, negative curvature, and zero curvature. These correspond respectively to elliptic, hyperbolic, and Euclidean (flat) geometries.

For nonmathematicians, non-Euclidean geometry seems abstract and counterintuitive. After all, the common meaning of parallel involves pairs of lines that never meet. If you try to parallel-park and veer into the next car, you can't ask the police for a non-Euclidean exemption. The triangles most children learn about in school are flat and their angles add up to 180 degrees. Why make geometry more complicated by changing its basic precepts?

As his ideas developed, but before they matured into his general theory of relativity, Einstein would wonder this himself. The geometry primer so pivotal to his early education firmly grounded him in the Euclidean tradition. He discussed his ideas with a family friend and medical student, Max Talmey (originally Talmud), who often visited. Talmey was struck by the depth of such a young boy's thoughts on mathematics, nature, and other subjects.

Einstein wouldn't learn about the non-Euclidean variety until his university years. Still clinging mentally to his childhood geometry book, he would initially dismiss it as something unimportant to science. It wasn't until much later that, thanks to the influence of his university friend Marcel Grossmann, he would come to see the importance of non-Euclidean geometry. By introducing non-Euclidean geometry to theoretical physics, Einstein would transform the field in extraordinary ways.[3] The twelve-year-old clutching the geometry book would have no way of knowing that his very hands would someday rewrite physical laws in a way that made the book obsolete.

Atoms in Motion

Vienna in the late 1890s was the home of raging debates in fundamental science. While Schrödinger was in the midst of his schooling, first through private tutoring and then, starting in 1898, at the prestigious Akademisches Gymnasium, two of the key figures who would help shape his intellectual life, Ludwig Boltzmann and Ernst Mach, were engaged in a heated argument about the reality of atoms.

When Boltzmann was appointed to the chair of theoretical physics at the University of Vienna in 1894, he had already made a name for himself as one of the founders of statistical mechanics (known then as kinetic theory), a field of physics that connects the behavior of tiny particles with large-scale thermodynamic effects such as temperature, volume, and pressure changes. To apply his techniques, he needed to assume that each gas is composed of enormous amounts of minuscule objects: atoms and molecules.

Boltzmann's achievements helped make thermal physics a hot item. Many young researchers were attracted to working with him in Vienna. Physicists Lise Meitner, Philipp Frank, and Paul Ehrenfest, who would all go on to successful careers, benefited from his supervision of their PhD research. Schrödinger was inspired by Boltzmann and as he approached university age hoped to work with him too.

Despite these accomplishments, Boltzmann's equilibrium was disturbed by the arrival of Mach. In 1895, Mach joined the faculty of the University of Vienna as chair for the philosophy of the inductive sciences. Pointing to the need for more experimental proof, Mach took a principled stand against atomism and Boltzmann's theories. Thermodynamics should be based upon what is perceived and directly measured, such as heat flow, he argued. He drew from a philosophical framework called positivism that rejects abstract knowledge and insists upon empirical evidence to support all propositions. Equating belief in atoms with religious faith, he preferred to stand on the side of what he saw as scientific rigor and the direct evidence of the senses.

"If belief in the reality of atoms is so important," Mach wrote, "I cut myself off from the physicist's mode of thinking, I do not wish to be a true physicist, I renounce all scientific respect—in short: I decline with thanks the communion of the faithful. I prefer freedom of thought."[4]

Mach did not aim his barbed logic at just Boltzmann. He targeted even the most venerated physicists whenever he saw their positions as divorced from the evidence of the senses. Daringly, he criticized one of the basic tenets of Newtonian mechanics, the notion of judging inertial states (at rest or at a constant velocity) by their relationship to an abstract framework called "absolute space." By that time, Newton had gained almost saintly status, particularly in Great Britain. Yet Newton's concept of inertia was built on an abstraction—exactly the kind of science Mach found suspect.

Mach's argument against Newton's definition of inertia referred to a thought experiment involving a rotating bucket that Newton had concocted to demonstrate the need for absolute space. Here's the gist of the argument: Imagine hanging a bucket filled almost to the brim with water on a rope tied to a tree. Now twirl the bucket carefully around and around until the rope is all twisted up. Hold the bucket, wait until the water within it has settled and has a flat surface, then let it go. The bucket will start to spin around on its own. If you look down within it, you'll see the water slosh around too as it forms a vortex, its surface becoming increasingly concave. That's because inertia makes the water try to escape. Since it can't leave the bucket, its outer edge rises. If you look at the inside of the bucket itself, ignoring its exterior, you might wonder why the water had a concave surface. Relative to the bucket, the water would seem to be perfectly still. Only by reference to an outside framework—which Newton called absolute space—would the concavity make sense. The water's rotation relative to absolute space, Newton asserted, remolded its surface.

Mach begged to differ, arguing that there was no empirical evidence for absolute space. More likely, he said, there was a pull on the water from unaccounted sources, such as the aggregate influence of distant stars. Just as the moon's tug causes the tides, perhaps the combined pull of the stars somehow causes inertia. Einstein would later dub this idea "Mach's principle." It would inspire him as he developed relativity.

Mach's critique of Newton stimulated a rethinking of classical mechanics that would spur Einstein and other physicists to consider alternatives. Mach's notion that science must offer perceptible evidence and eschew hidden mechanisms greatly influenced Schrödinger, who delved into his writings with gusto. Yet his attacks on Boltzmann's atomism may have taken a personal toll. Prone to intense mood swings and suffering from declining health, Boltzmann hanged himself in September 1906, while on vacation in Trieste with his family.

University Days

By cruel fate, Boltzmann's suicide happened just a few months before Schrödinger began his studies at the University of Vienna in the winter of 1906/1907. Schrödinger had graduated from the Akademisches Gymnasium as a star pupil in mathematics and physics, his favorite subjects. First in his class, he could have majored in practically anything, but his passion was in the equations describing the physical world. He was keen to pursue theoretical physics at university, and Boltzmann would have been an extraordinary mentor. Alas, he entered university at a somber time, with a cloud hanging over the physics program.

"The old Vienna institute, which had just mourned the tragic loss of Ludwig Boltzmann, . . . provided me with a direct insight into the ideas of that powerful mind," Schrödinger recalled. "His world of ideas may be called my first love in science. No other personality has since thus enraptured me or will do so in the future."[5]

Schrödinger was stirred by Boltzmann's bravery in attacking fundamental questions. With his atomic building blocks, Boltzmann was unafraid to construct principles governing the thermal behavior of the entire universe. Inspired by his example, later in life Schrödinger would be similarly ambitious in trying to identify a basic theory encompassing all of the natural forces.

Boltzmann's replacement for the university's theoretical physics chair was one of his former students and an excellent theoretician, Friedrich "Fritz" Hasenöhrl. Hasenöhrl made his name in the study of electromagnetic radiation emitted by moving objects and found a relationship between energy and mass (though he was off by a factor) even before Einstein's famous equation.[6] He was friendly and welcoming to students. Given that he couldn't study heat theory and statistical mechanics under Boltzmann, Schrödinger was privileged to study those subjects and others, such as optics, under Boltzmann's well-trained successor. Hasenöhrl was by all reports an outstanding teacher. Inspired by Hasenöhrl's teachings and Boltzmann's achievements, Schrödinger hoped to carve out his own path of discovery in theoretical physics.

Schrödinger quickly developed an excellent reputation as a student. Hans Thirring, a fellow physics student who became a lifelong friend, recalled sitting in a math seminar, seeing a fair-haired youth enter the room, and hearing another student who knew him from his school days remark with awe, "Oh, this is Schrödinger!"[7]

Despite his theoretical interests, the major thrust of Schrödinger's university research was experimental work guided by Franz Exner. Schrödinger would receive his doctorate under Exner's supervision. Exner was interested in the many manifestations of electricity, including its production in the atmosphere and through certain chemical processes. He also explored the science of light and color and investigated radioactivity. Schrödinger's doctoral dissertation was entitled "On the Conduction of Electricity on the Surface of Insulators in Moist Air." It was a very practical thesis, concerned with the problem of insulating devices used for physical measurements from the electrical effects of moisture. The future theorist started his career getting his hands dirty, working in a small lab attaching electrodes to samples of amber, paraffin, and other insulating materials and measuring the currents flowing through them. He received his doctorate in 1910 and his *Habilitation* (the highest academic degree in the Austrian educational system, allowing one to teach), based on a theoretical problem related to atomic behavior and magnetism, in 1914.

It would not be until many years later that Schrödinger and Einstein would start to explore the unification of gravitation and electromagnetism. Yet, strangely enough, a 1910 letter from the ailing Mach that ended up in Schrödinger's hands would anticipate those efforts. Although Mach had retired, his mind was still actively pursuing deep questions about nature. He had started to wonder about commonalities in the inverse-squared laws of gravity and electricity, pondered if these forces could be unified, and inquired who at the university might be able to answer his questions. In particular, Mach wanted someone knowledgeable to assess the theories of controversial German physicist Paul Gerber. Mach's query was passed along to Schrödinger, who found Gerber's writings hard to follow. Nevertheless, the exchange represented an indirect encounter between Schrödinger and one of his intellectual heroes, Mach, and was a harbinger of Schrödinger's theoretical work to come. Moreover, that he was chosen as the one to respond to Mach was a sign of the high regard in which Schrödinger was held at the university. Still just in his mid-twenties, Schrödinger was starting to make a name for himself.

Racing After Light

While Schrödinger never had a chance to work with Boltzmann, he nevertheless found much meaning and achievement in his studies. He was clearly a star student. Einstein's university life was marked by a different cause of disappointment: he did not have the opportunity to study the deep theoretical questions he was really interested in. Consequently, he did not take all of his classes as seriously as he should have, particularly his math courses, as they didn't seem relevant to his intellectual passions. Nonetheless, the personal connections he made would prove pivotal to his intellectual growth.

Einstein's transitions from high school to university and then from university to an academic career were much bumpier than Schrödinger's. In 1893, Einstein's father lost his electrification contract with the city of Munich. The following year he dissolved his firm and decided to move the family to Milan, Italy, in search of work. Einstein was still completing his schooling at the Luitpold Gymnasium and needed to remain in Munich without his family. Several months later, Einstein decided that it would be best to leave Germany too, applied successfully for early release from his school, and obtained permission to take university entrance exams early. The university he chose was the Swiss Federal Institute of Technology in Zurich, known by its Swiss acronym ETH (Eidgenössische Technische Hochschule).

It was around that time, at the age of sixteen, that Einstein had an unusual vision—imagining himself chasing a light wave and trying to catch up with it. If he could travel at the speed of light, he wondered if he would see the wave just oscillating in place. After all, if you run alongside a bicycle, it looks like it is standing still. As Newton pointed out, moving at a constant speed and being at rest are both inertial frameworks that share identical laws of motion. Thus if two things are traveling together at the same velocity, they should appear to each other exactly as if they were each at rest. However, Maxwell's equations of electromagnetism make no reference to whether an observer is moving or standing still. According to those laws, light should travel through space at always the same speed. Einstein realized that Newton's and Maxwell's predictions blatantly contradicted each other. Only one of them could be right—but which?

The idea that the speed of light in a vacuum was constant—or even that light could travel through pure emptiness—was not widely accepted at the time Einstein was pondering this question. Many physicists of the day believed that light moved through an invisible substance called the "luminiferous aether," or just "aether" for short. Earth's motion relative to the aether should thereby be detectable. However, a well-known experiment in 1887 by American physicists Albert Michelson and Edward Morley had failed to detect such an effect. To try to reconcile light's behavior with Newton's laws of mechanics, Irish physicist George FitzGerald and, independently, Dutch physicist Henrik Lorentz suggested that fast-moving objects compress along their directions of motion. Such a shortening, called the Lorentz-FitzGerald contraction, would squash the instruments of the Michelson-Morley experiment in such a way to make it appear that the speed of light was always constant. Unaware at the time of the Michelson-Morley experiment, Einstein considered the question independently without invoking aether. He somehow had a hunch, even before he read Mach, that Newtonian physics was ailing and required radical surgery.

Remarkably, given his later reputation as the world's foremost genius, Einstein failed the ETH entrance exams the first time he took them. Perhaps this failure was one of the sources of the folk myth that he failed math in school. Actually, it was his French-language essay that proved to be his weakest point. He beefed up his skills by attending a high school in Aarau, Switzerland, for one year. Daringly, he renounced his German citizenship, as if to sever all ties with his earlier life. Living without his parents, and, for the time being, stateless, he was a most unusual teenager. Fortunately, he passed the exams the second time around and was accepted to ETH at the virtually unprecedented age of seventeen.

Once enrolled at ETH, Einstein found that physics there was very old-fashioned, focused on traditional subjects such as mechanics, heat transfer, and optics. The Machian critique of Newton had not penetrated its hallowed halls. Maxwell's theory of electromagnetism was little in evidence. Einstein still thought about his light speed problem but would not find a solution within the university's curriculum.

Einstein's years at ETH would correspond to an extraordinary time for physics. While the Mach-Boltzmann debate about atomism was raging in Vienna, in 1897 Cambridge physicist J. J. Thomson provided experimental proof for an elementary particle far smaller than an atom. His colleagues were dubious at first that something could be much tinier than supposedly indivisible things. Thomson dubbed the negatively charged particle a "corpuscle," but FitzGerald, following the suggestion of his uncle, Irish scientist George Stoney, gave it the name that stuck: "electron." In Paris, Henri Becquerel discovered radioactivity, exploring the properties of radioactive uranium along with his doctoral student Marie Curie and her husband, Pierre Curie. In 1898, the Curies identified radium, another radioactive element. All of these findings pointed to the complexity of atoms—a topic that would later engage Einstein, Schrödinger, and many other physicists of their generation. Yet at ETH, students were encouraged to stick to time-tested, practical physics. It would be a poor match for Einstein's yearnings for innovative explanations of natural phenomena.

Einstein was lucky to find a circle of friends who supported each other with their studies and off whom he could bounce ideas. One of his key sounding boards—whom he met outside of the university through their shared love of music—was a bright Swiss-Italian engineer named Michele Besso. Besso profoundly influenced Einstein's career by introducing him to the writings of Mach. Einstein and Besso would be dear friends for life.

Another steady companion was Marcel Grossmann, who was a whiz in higher mathematics. He took excellent notes in math classes, which Einstein came to rely on whenever he decided to skip a class—which was often. Later Grossmann would become a math professor at ETH and help Einstein develop the mathematical framework behind the general theory of relativity.

Given the prestige of his instructors at ETH, Einstein should have paid closer attention to math. One of his professors was Hermann Minkowski, who would later help reframe Einstein's theory of special relativity in a more elegant, useful way. Minkowski was born in Lithuania and educated at the prestigious University of Königsberg. He was one of the few professors at ETH with the skills for injecting vital higher mathematics into the body of theoretical physics. Ironically, given their mutual fate, at that point he thought little of his distracted student. Noting with great concern how many times Einstein missed his class, Minkowski called him a "lazy dog."

Einstein later justified his lack of attention to math by noting: "It was not clear to me as a young student that access to a more profound knowledge of the basic principles of physics depend on the most intricate mathematical methods. That dawned on me only gradually after years of independent scientific work."[8] Ideally, Einstein should have focused more on the skills he would need for theoretical physics. However, there was good reason for him to be distracted from his classes in general. By his second year in university, he had fallen in love with his only female classmate, a young Serbian named Mileva Marić. Their fiery passion manifested itself in flirtatious letters and love poems that became public long after Einstein's death. Einstein's relationship took on a bohemian character as he sought a connection with her based on true equality, free love, and complete support for each other's intellect and goals. Envisioning him with young women more like their family in terms of class, values, and ethnic background, Einstein's mother strongly disapproved of the relationship. Nevertheless, their heated romance persisted—the family obstacles recasting their ardor as a fervent revolutionary struggle.

In his third year at ETH, Einstein took several physics courses but failed to make a good impression. For one such class, called "Physical Exercises for Beginners," his attendance was so poor that its instructor, Professor Jean Pernet, scolded him and gave him the minimum possible grade. Another course, on the subject of heat and taught by Professor Heinrich Friedrich Weber, disregarded the advances made by Boltzmann and others. Einstein decided to study Boltzmann on his own. The highlight of the year's curriculum for him was having the opportunity to work in Weber's Electrotechnical Laboratory, where he became acquainted with state-of-the-art equipment. Despite his earnest attempts to impress Weber, the practical-minded professor had little patience for the disheveled, idealistic youth.

To little avail, Einstein tried to convey to Weber his interest in solving the speed of light problem. He proposed using Weber's lab to try to measure Earth's motion through the aether, not realizing that Michelson and Morley had completed such an experiment some years earlier. Unsurprisingly, given his complete lack of interest in Maxwell's electromagnetic theories and other recent advances, Weber was skeptical and didn't support such a retread. It didn't help Einstein's reputation when, ignoring written instructions, he damaged his hand by causing an explosion in Pernet's lab. As his academic studies at ETH drew to a close, his performance had given the faculty little reason for confidence. Upon completing his final exams and receiving his diploma as a teacher of math and physics, he tried unsuccessfully to get a position at ETH as a research assistant. To his great shock and dismay, none of the professors—mathematicians and physicists alike—wanted to take him on. "I was suddenly abandoned by everyone," Einstein painfully recalled, "standing at a loss on the threshold of life."[9]

To make matters worse, he watched as almost all his classmates, including his close friend Grossmann, found postgraduate positions at ETH. Mileva was an exception; she had performed poorly in her final exams and was held back from graduating. Lacking the support of any of his professors, he had nowhere to go. Only a series of miracles could rescue his career.

The Road to Miracles

As bells chimed for the turn of the century, the physics community was divided about the state of the profession. Older physicists, wrapped snugly in the cloak of Newton, generally saw the field as close to completion, with only a few loose ends to tie together. Younger physicists, donning lab smocks for hands-on exploration of radiative and electromagnetic effects, weren't so smug about strange new phenomena that defied explanation—from invisible X-rays to glowing radium.

On April 27, 1900, British scientist Lord Kelvin (William Thomson) delivered a speech, "Nineteenth-Century Clouds over the Dynamical Theory of Heat and Light," that delineated the two main problems he believed to be obscuring the way forward in physics. Once these "clouds" cleared up, he assumed, physics would have a sunny future. Little did Kelvin know that he had pinpointed the twin issues that would compel revolutionary changes in the field.

Kelvin's first "cloud" was the issue of light's motion through space, centering on the question of why the Michelson-Morley experiment failed to detect the aether. While Lorentz and others had made suggestions, the issue remained unresolved. Kelvin hoped for a more satisfactory explanation.

His second nebulous quandary involved the emission of radiation from a blackbody. Theoretical models simply didn't match up to known experimental results. Something seemed wrong with the assumptions being made.

A blackbody is a perfect absorber of light. Think of a box coated with jet-black paint that soaks in every drop of light shined on it. A blackbody can also emit light, releasing radiation at various wavelengths. Some of these wavelengths correspond to visible colors, ranging from short-wavelength violet to long-wavelength red. Other wavelengths represent invisible forms of light, including the shorter-than-violet form called ultraviolet and the longer-than-red form called infrared. We now know that the electromagnetic spectrum ranges in wavelength from ultra-tiny gamma rays to sizable radio waves.

As nineteenth-century scientists noted, the distribution of radiation among different wavelengths depends on the temperature of the body emitting the light. The hotter the object, the more its peak wavelength shifts toward the shorter end. We can see this when things burn: hotter fires glow with blue flames, while fires that are not as hot glow orange or red. Humans and most living things are cool enough to give off light mainly in the infrared range.

Lord Rayleigh (John William Strutt), the accomplished successor to Maxwell at Cambridge, carefully applied wave theory and statistical mechanics to the study of blackbody radiation. Computing how many peaks of a particular wavelength could fit into a cavity, he developed a distribution formula that favored shorter wavelengths. His logic made sense: you can fit more shorter things into a box than longer things. He published his analysis in 1900.

The problem with Rayleigh's model is that it predicted a surge of low-wavelength, high-frequency radiation each time a blackbody releases light. (Frequency is light's rate of oscillation; the shorter its wavelength, the higher its frequency.) Instead of glowing orange, red, or blue, a fire, by that line of reasoning, should always be invisible. Heat up a dark coffee mug, leave it on a table, and it could blast you with skin-burning ultraviolet or even hazardous X-rays instead of warm, friendly infrared. Ehrenfest would later dub this problem the "ultraviolet catastrophe."

In a rare case of a seemingly intractable problem finding a speedy solution, later that same year German physicist Max Planck proposed the idea that energy comes in tiny packages or "quanta." These are whole-number multiples of the frequency times a tiny number known as Planck's constant. Planck was not specifically addressing Rayleigh's calculation but rather the general question of how blackbodies radiate. By restricting light's energy to finite bundles, with amounts proportional to frequency, Planck found that he could skew the distribution toward more moderate frequencies and wavelengths. That's because there was more of an energy "cost" for higher frequencies (short wavelengths) than lower frequencies (long wavelengths).

It is like filling a piggy bank with a pile of coins of various denominations, including quarters and pennies. Since quarters are larger than pennies, fewer of the former and more of the latter would fit. Therefore, one would expect mostly pennies in the bank. However, if the coins were from a pricey collection in which the pennies were rarer and more expensive than the quarters, there would likely be fewer pennies available. Thus the higher cost of the pennies would balance their small size, allowing for a more even mix within the piggy bank. Similarly, in

Planck's model, the greater energy cost of high-frequency quanta would balance their smaller wavelengths, ensuring a more even distribution matching physical reality.

Planck intended the idea of discrete quanta to be a mathematical tool rather than a physical constraint. However, the years to come would establish the quantum idea as pivotal to a radical remaking of physics. Thanks to his advancement of the photoelectric effect during his *annus mirabilis* or "miracle year" of 1905, Einstein would play a strong role in that development.

Einstein's miracle year was preceded by a period of intense intellectual effort. He somehow managed to complete these groundbreaking calculations during a time of financial hardship. "His environment betrayed a great deal of poverty," recalled Max Talmey. "He lived in a small, poorly furnished room. He . . . was struggling hard for a living."[10]

Lacking an academic position, Albert supported himself and Mileva at first by tutoring and then by working at a patent office in Bern as a "third-class technical expert"—a job he obtained through the help of Grossmann's father, who knew the director. While assessing blueprints for new inventions and deciding whether they were workable and original, he managed to steal time for the pursuit of deep questions in physics. Because of his efficiency, he soon found that he could fulfill his job responsibilities each day through a few hours of effort, leaving the rest of the time for his own research.

Part of the financial pressure on Einstein that led him to pursue the patent office position was that Mileva had become pregnant. Though he tried to reassure her that everything would work out, it was not a happy period for her. Her own scientific career was in shambles, as she had failed her final exams for the second time. Albert had promised to be supportive but instead had his nose stuck in his own work.

In late 1901, Mileva returned alone to her native city of Novi Sad. There, at her parents' house, she gave birth to their daughter, Lieserl, in January 1902. The rest of Lieserl's life is lost to history; some historians conjecture that she was adopted by a Serbian family and died at a young age. Einstein likely never met his only daughter, the existence of whom he kept secret from his parents, other family members, and friends. Only a hidden box of letters, opened by historians well after his death, revealed her existence.

Mileva returned to Bern, and the two got married in January 1903. Later that year they moved to an apartment on Bern's main street, Kramgasse, near its famous clock tower. They would have two more children together, Hans Albert (born in 1904) and Eduard (born in 1910). Instead of pursuing a physics career, she supported his while taking care of the children and managing the household. Her dreams unfulfilled and their marriage strained, she slumped into a humdrum existence filled with depression. On life's seesaw she would sink and he would soar.

Largely free of domestic responsibilities and finding his job not very challenging, Einstein found time to discuss philosophy with a group of friends he had met shortly after he first arrived in Bern. Fashioning themselves after the ancient Greeks, they called themselves the "Olympia Academy." The founding member was Maurice Solovine, a student from Romania who was interested in an eclectic range of subjects. He had originally responded to a tutoring ad Einstein had posted, but the relationship quickly became a friendship. The other steady member of the group was mathematician Conrad Habicht. Meeting regularly, they discussed readings from works by Mach, Poincaré, Spinoza, and many others. Their lively debates helped shape Einstein's thinking as he developed his pivotal contributions to human knowledge.

With the hope of returning to academic life, in early 1905 Einstein completed a doctoral dissertation for the University of Zurich. He had developed a formula to determine the dimensions of particles in a solution by measuring the fluid's viscosity (resistance to flow). Nothing in this practical work would hint at the explosion of ideas about to be ignited.

In the spring of that year, Einstein took aim. Staring classical physics straight in the face, he lit the fuses and launched his grenades. He submitted four papers to a prestigious journal, *Annalen der Physik*. One was a version of his dissertation. The other three articles—addressing the photoelectric effect, Brownian motion, and the special theory of relativity—would shake the foundations of physical science.

Einstein's paper on the photoelectric effect cemented Planck's quantum idea by making it tangible and eminently measurable. It considers what would happen if a researcher shines a light on a metal, supplying enough energy to release an electron. If light were purely a wave, theory suggested, its amount of energy would depend mainly on its brightness. Thus a bright flash of red light would necessarily convey more energy than a dim exposure to ultraviolet light. Brightness can be cranked up or down in a continuous way. So if it were the major factor, light's energy could be set to any amount. As the light's energy jostled the electron, it would flee the metal with a speed related to the brightness—the brighter the light, the faster the electron.

However, Einstein took the radical step of suggesting that in some cases light acts like a particle, later called a photon. Each photon carries a discrete package of energy that is proportional to the light's frequency. Thus higher-frequency light sources emit photons that are more energetic than lower-frequency sources do. For example, blue light conveys more energy per photon than red light. Consequently, shining a higher-frequency light on a metal offers more of a chance of freeing an electron and having it move faster than does shining a lower-frequency light on it. The speeds of the emitted electrons correlate beautifully with the frequencies of the light cast on the metal—a result that has been duplicated countless times in physics labs around the world.

By identifying the photoelectric effect—showing that electrons emit and absorb light in discrete quanta—Einstein provided an important clue as to the workings of atoms. These insights would prove critical to Danish physicist Niels Bohr in his development of an atomic model less than a decade later. Bohr would show that by gobbling up a photon, an electron rises to a higher energy state, and by emitting one, it sinks to a lower state.

Einstein would certainly be well known if the photoelectric effect was his major contribution that year. Indeed, when he received the 1921 Nobel Prize in Physics, that accomplishment was the one cited. But that result was just the overture of his grand symphony of scientific revelations.

Another major paper Einstein published in 1905 concerned a phenomenon called Brownian motion, named after the Scottish botanist Robert Brown, involving tiny random fluctuations of small particles. In 1827, Brown had observed the agitated motion of particles found in pollen grains immersed in water. He failed to find a credible explanation for their erratic behavior. Building upon his doctoral thesis, Einstein decided to model the movements of particles bashed around by water molecules and discovered precisely the kind of haphazard jig seen by Brown. By explaining Brownian motion as the zigzag result of myriad particle collisions, Einstein furnished important evidence for the existence of atoms.

Arguably Einstein's most important contribution during his *annus mirabilis* was the special theory of relativity. He finally addressed the question that had occupied his late teenage years about chasing a light wave. No matter how fast you go and how hard you try, he concluded, light waves were uncatchable. Nothing made of matter could ever reach the speed of light.

Today science is accustomed to the idea of a universal speed limit, but back then the notion was virtually unthinkable. Classical physics, as described by Newton and taught for centuries as immutable laws, maintains that relative speeds are simply additive. Thus if you were skateboarding west at a particular speed relative to the deck of a ship and that ship was moving west at a certain speed relative to the ocean, the two speeds would add up. Your speed relative to the ocean would be the sum of the speed you manage and the speed the ship puts out. If the ship somehow zipped across the ocean at two-thirds the speed of light and you could muster the same astonishing rate on your skateboard, you could outpace the speed of light easily.

In the age of Edison, power seemed limited only by the imagination. Given that electricity could illuminate cities, propel streetcars and trains, and power factories, surely enough energy could be found in the world to accelerate anything up to any velocity. If a single battery could make something move at a certain pace, nothing in the laws of motion ruled out making it move billions of times faster through billions of batteries.

By taking Maxwell's equations of electromagnetism at face value, however, and disregarding any influence of the aether if it existed at all, Einstein argued that the speed of light in a vacuum is an absolute constant no matter who was measuring it. Voyagers racing at incredible speeds alongside a light beam would still observe it dashing away from them at the same velocity as if they were at rest. Therefore, like a desert mirage, no matter how fast someone tried to travel, light would remain elusive.

To reconcile the constancy of light speed with the concept of relative velocity, Einstein realized that he would need to tear up parts of the Newtonian rulebook. He decided to discard the concepts of absolute time and space (the latter so disliked by Mach) and replace them with more malleable notions. If, he reasoned, moving observers saw clocks ticking slower and yardsticks shrinking in the direction of motion, light's speed could maintain the same value. These two ideas—time dilation and length contraction—brought together Maxwell's theory with an amended theory of motion, banishing one of Kelvin's clouds and bringing on a sunny future.

Time dilation involves a discrepancy between the "proper time" of an observer moving along with something under study and the relative time of a second observer traveling at a different constant speed compared to the first. For example, suppose that the first observer is a passenger on a spaceship moving close to the speed of light. For that passenger, the time on the ship's clock would be the proper time. However, if a second observer—the passenger's sister on Earth, let's say—somehow managed to see the clock (using a superpowerful telescope aimed at the ship's large windows)—she would measure it to be moving slower.

To understand the reason behind the discrepancy, imagine that the passenger spends his time playing a kind of ping-pong game with a beam of light, In this pastime, he bounces the light off a mirror on the ceiling—shining it straight up, seeing it reflect straight down, and timing how long this takes. Peering remotely at her brother's hobby while seeing his ship soar through space, the sister would observe the light to be moving in a zigzag, rather than straight up and down. The horizontal movement of the spaceship combined with the rising and falling of the light would produce a kind of inverted V shape. If she assumed that the speed of light was constant, she would conclude that the light traveled a longer distance and thereby took a longer time compared to what her brother would measure. Thus she would see time on the ship as passing more slowly than her brother would report.

Relativistic length contraction is a variant of the Lorentz-FitzGerald contraction, involving the compression of space itself in the direction of motion, rather than a squashing of matter. While those observers moving in tandem with an object would experience its proper length, others traveling at a different constant speed would measure its length to be shorter along the direction of its movement.

To help comprehend this idea, imagine that the space passenger decides to play "light ping-pong" against the front wall of the ship (in the direction of the ship's motion) instead of the ceiling. Placing a mirror on that wall and aiming his light generator toward it, he bounces the beam horizontally back and forth. Multiplying the time elapsed by the speed of light, he determines the total length of the path the beam traveled. On Earth, his sister aims her superpowerful telescope at the ship and measures the length of the beam's path as well. Because the ship is racing in the same direction as the light beam (before it bounces), she records the time required for the beam to bounce back and forth as being quicker than her brother's estimate. Consequently, she observes its path length to be shorter.

A follow-up paper about the special theory of relativity showed what would happen to mass at high speeds. Einstein proposed that relativistic mass was a form of energy, linked by the now-famous equation $E = mc^2$. An object starts with a certain quantity of rest mass—its birthright, so to speak. By moving faster, it accrues additional mass linked to its energy of motion. The closer it gets to the speed of light, the higher its mass. To actually reach the speed of light an object would need to convert an infinite amount of energy to mass—an impossibility. Thus for material bodies, the speed of light is unreachable (unless the object is already at that speed).

The Union of Space and Time

After Einstein published his extraordinary results, the German scientific community began to take notice. It would be a while, however, before he would acquire international fame. One early supporter was the physicist Max von Laue, then an assistant to Planck in Berlin. In the summer of 1906, von Laue took the time to schedule a visit with Einstein at the patent office. He sat in the waiting room, eagerly awaiting the chance to meet the prodigious successor to Newton's throne.

"The young man who came to meet me," von Laue recalled, "made so unexpected an impression on me that I did not believe he could possibly be the father of the relativity theory. So I let him pass and only when he returned from the waiting room did we become acquainted."[11]

Von Laue did much to publicize Einstein's theory of relativity and explore its implications. He would write the first primer about relativity, published in 1911. Einstein greatly appreciated his friendship and support, which lasted throughout their lifetimes.

Another cheerleader for Einstein was Minkowski, whose view of his former student took a 180-degree turn. Astonished that his "lazy dog" student had solved the long-standing puzzle of the interpretation of Maxwell's equations, Minkowski decided to reframe the theory in a more mathematically rigorous way. By then, he had been appointed to a position in the mathematical mecca of Göttingen, where the influential logician and geometer David Hilbert had assumed Klein's mantle as the chief exponent of innovation in the field. In that center for all that is beyond Euclid, Minkowski was well placed to take advantage of groundbreaking geometric approaches.

Minkowski brilliantly determined that Einstein's theory would look far more elegant dressed up in the garments of four-dimensional geometry. He fashioned an alternative to Euclidean space that had two key differences. The first was that time (multiplied by the speed of light to get the units right) was included as a fourth dimension. Length, breadth, and height were joined by duration as ways of describing nature. He called this amalgamation "spacetime."

The second change involved adding a negative term to the Pythagorean theorem used to determine distances. Its standard version, used for millennia to find the hypotenuses of right triangles, states that the sum of the squares of the sides of a right triangle equals the square of its hypotenuse. For example, in a right triangle where the sides are 3, 4, and 5 units, $3^2 + 4^2 = 5^2$. Minkowski modified this to include time by stating that the sum of squares of spatial distances minus the square of the fourth coordinate (time multiplied by the speed of light) equals the square of the "spacetime interval." The spacetime interval is the shortest path length in four dimensions, a generalization of spatial distance that takes both space and time into account. It represents the proximity of two events—things happening in different places and different times—by measuring the minimal four-dimensional route linking one to the other.

Knowing the spacetime interval between events tells you whether or not they are causally connected—meaning one could affect the other. If the spacetime interval is zero, called "lightlike," or negative, called "timelike," the earlier event might influence the later one. On the other hand, if the spacetime interval is positive, called "spacelike," no causal communication is possible, as that would require a faster-than-light signal. So, if an actress dressed in a certain style during the 2016 Oscars ceremony and a Proxima Centaurian four light-years away adopted the same style in 2017, she couldn't be accused of being a copycat, as the interval between them would be spacelike, with no possibility of causal communication. A signal would have taken a minimum of four years, not one year. The Centaurian fashion statement would merely be cosmic coincidence.

By framing the special theory of relativity as a four-dimensional theory set in spacetime, Minkowski showed how time dilation and length contraction could be construed as rotations that transform space into time. To see how such rotations occur, we can think of the spacetime interval as something like a weather vane, with north representing time and east representing space. Switching between two different perspectives is like a turning of the vane from east-northeast to north-northeast—taking away some of its easterly component and replacing it with a more northerly component. Similarly, a spin of the spacetime interval might take away some of the spatial distance between two events while increasing their separation in time.

Minkowski triumphantly announced his findings at the Eightieth Assembly of German Natural Scientists and Physicians in Cologne, emphasizing their revolutionary nature: "The views of space and time which I wish to lay before you have sprung from the soil of experimental physics and therein lies their strength. They are radical. Henceforth space by itself and time by itself are doomed to fade away into mere shadows, and only a kind of union of the two will preserve an independent identity."[12]

While Einstein at first dismissed Minkowski's reformulation of his theory as too intellectually demanding, within a few years its brilliance would soak in. It would have a profound influence on his mode of thinking, helping him realize the critical importance of higher-level mathematics in advancing physics.

In 1908, the same year as Minkowski's announcement, Einstein received his *Habilitation* and began to teach at the University of Bern. The following year he was appointed to a position at the University of Zurich. There he began to plan out a sequel to the special theory: a comprehensive theory of gravitation called the general theory of relativity. To do so, he would need to rethink his views about higher mathematics.

It was time to move beyond the limitations of childhood. The worn geometry book with its planar Euclidean mathematics had served the young Einstein well, but to advance his theory he would need to embrace non-Euclidean geometries and a fourth dimension. Einstein's progression would in turn inspire Schrödinger, whose childhood fascination with astronomy—exemplified by the "dance of the planets" with his aunt—would deepen into an interest in the relativistic approach to gravitation. The two would find themselves grappling with theoretical questions in the midst of a turbulent age for Europe—war, economical collapse, political turmoil, and more war.

CHAPTER TWO

The Crucible of Gravity

A patriot fiddler-composer of Luton
Wrote a funeral march which he played with the mute on
To record, as he said, that a Jewish-Swiss-Teuton
Had partially scrapped the Principia of Newton.

—LIMERICK PUBLISHED in *Punch,* 1919[1]

While Newton's theory of gravitation was elegant in its simplicity, Einstein found it fundamentally unattractive. It treated gravity as an instant, invisible connection between two distant masses. Unseen threads of gravitational force somehow guided the celestial bodies through space. Agreeing with Mach that nature should be measurable and observable, Einstein sought a deeper explanation.

Moreover, the special theory of relativity set an upper limit to the swiftness of causal communication: the speed of light. Newton's theory didn't obey such a rule. If the Sun disappeared, the theory predicted, Earth would immediately follow a straight path through space—even before the last rays of sunlight arrived. How would it know to do so even before the Sun's absence could be communicated? Einstein realized that gravitation thereby needed to be recast in the language of relativity.

As an ardent appreciator of Maxwell's approach to electromagnetism, which was based on the concept of fields, Einstein similarly wanted to compose a field theory of gravitation. A field is a mapping of a force's potential impact with particular values for each point in space. The field's strength at a certain location helps determine how much force a particle placed there would experience. An electric field, for example, sets out how much electric force an electron, proton, or other charged body would feel at any given place. A magnetic field does the same for magnetic force.

Consider, for instance, a field that represents the strength and direction of waves throughout the ocean. A hapless sailor who found himself in a place where the field was exceptionally strong might witness his vessel rocked by overwhelming forces, wresting him off course. Even if he didn't know the source of the powerful waves—such as an undersea earthquake—he would experience its terrifying impact firsthand. Thus while the cause of the disturbance might be distant, the field would act as a conduit and the impact would be local. Seeing marked similarities between electromagnetism and gravitation, such as the fact that their strengths drop off with the square of the distance between objects, Einstein set out in the early 1910s to identify gravity's field equations. The result would be his masterly general theory of relativity. By making an analogy between the forces he would set the stage for his eventual efforts to try to unify them.

In the middle of his struggle, Einstein would come to Vienna and deliver a progress report. His stirring conference talk would inspire young Schrödinger, then in his mid-twenties, to turn from practical topics, such as the measurable properties of light and radiation, to more fundamental questions, such as the riddle of gravitation and the properties of the universe itself. The connections Einstein drew between electromagnetism and gravitation would plant the seeds for Schrödinger's interest, much later in life, in finding a unified theory of the natural forces. The Vienna conference of 1913 would prove a turning point in his career. With Einstein as his role model, nothing in the cosmos would seem beyond the grasp of his clever mind.

Twilight of an Empire

The shining imperial capital of the Austro-Hungarian Empire was about to lose its luster. Its central fire would soon be extinguished and its satellite subjects released like embers to the wind. Its extinguishing would arrive as quickly and completely as a solar eclipse. Yet not all was dire. In such moments of darkness, stars that would never be seen by day have a glorious chance to sparkle.

The Habsburg city was throwing a party—a festive meeting of minds that, as it turned out, would serve as a farewell to Vienna's golden age. Invited were thousands of the finest German-speaking scientists in Europe. From Prague to Budapest and from Berlin to Zurich, young and old congregated to seek word on astonishing new theories about particles, atoms, light, electricity, statistical physics, and other topics. There were notable absences: Planck and Arnold Sommerfeld, the esteemed director of Munich's Physics Institute, did not attend. However, enthusiasm for new findings in physics made the last waltz of Austro-Hungarian physics a gala to remember.

No luxury was spared for that year's Assembly of German Natural Scientists and Physicians (the same group Minkowski had addressed half a decade earlier in Cologne). It met from September 21 to 28, 1913, in the spanking-new headquarters of the University of Vienna's Physical Institute, near the little street of Boltzmanngasse. Franz Exner had insisted on the construction of the new building as a condition for remaining director. After sessions at the building's grand lecture hall, the more than seven thousand conference members had the option of attending a sumptuous reception hosted by the Imperial Court, a banquet held by the Vienna city government, and a party graciously arranged by the Viennese physicists themselves. Surely no one complained about being underfed.

Among the topics of discussion, radiation and atomic physics were all the rage. One of the speakers was German physicist Hans Geiger, inventor of the Geiger counter (proposed in rudimentary form in 1908) and a former coworker of famed New Zealand–born physicist Ernest Rutherford. In 1909, under Rutherford's supervision at the University of Manchester, Geiger and Ernest Marsden had conducted an artful experiment designed to probe the atom. Bombarding gold foil with alpha particles (a type of radiation identical to helium ions), they discovered that almost all the particles passed unhindered through the foil. However, a small fraction bounced back at sharp angles, like superballs ricocheting off a concrete wall. From these unexpected results Rutherford deduced that atoms are mainly empty space but have tiny positively charged cores called nuclei. His 1911 rudimentary model of the atom as something akin to the solar system, with negative electrons orbiting a positive nucleus, fundamentally altered the concept of the atom. No longer could atoms be thought of as indivisible and solid, like tiny marbles; rather, they were intricate bodies principally composed of pure emptiness. Geiger's talk at the meeting focused on practical ways of detecting alpha and beta particles (the latter another type of radiation later identified as electrons).

As a young researcher working at Exner's Physical Institute and the nearby Institute for Radium Research, Schrödinger had become interested in detecting radiation too. One month earlier, Schrödinger had gone to the village of Seeham, on Lake Obertrumer near Salzburg, to record the amount of a radium decay product, called radium-A, in the atmosphere. Taking almost two hundred measurements using collection tubes and an electrometer, he calculated how the atmospheric content of radium-A changed over time. Curiously, he showed that even at its peak, radium-A accounts for only a fraction of atmospheric radiation. Based on his and other readings, many scientists had concluded that other sources, such as gamma rays, must supply the rest. Researchers had begun to probe possible causes of the extra radiation.

Relevant to his work and ideally located in his home city, the late September conference was just perfect for him. He could listen to talks describing the latest findings about radioactivity, the atomic nucleus, and related topics. One such talk, by astrophysicist Werner Kolhörster of Halle, Germany, described balloon flights miles above the ground equipped with radiation detection equipment. Confirming earlier results by Austrian physicist Victor Hess, he reported how "penetrating radiation," apparently from extraterrestrial sources, increases significantly at high altitudes. We now call this radiation from beyond Earth "cosmic rays." Because of this confirmation, science historians Jagdish Mehra and Helmut Rechenberg have suggested that the conference was the "birthday of . . . cosmic radiation."[2]

At the conference, many of the attendees, including Einstein, learned for the first time about Bohr's remarkable theory of atomic structure proposed earlier that year. Einstein thought that Bohr's achievement was "one of the greatest discoveries."[3] While none of the research talks specifically mentioned the Bohr model, word of the triumph arrived informally via the personal account of Hungarian physicist George de Hevesy, who had witnessed its development firsthand. De Hevesy had been in Manchester in 1912 when Bohr was a visiting postdoctoral scholar working with Rutherford. He saw how Bohr and Rutherford's collaborative efforts had proven most fruitful in advancing atomic theory. Then de Hevesy was appointed to the Institute for Radium Research in Vienna, where he was in an ideal position to convey the exciting news about Bohr's work to interested conference participants.

Bohr took Rutherford's planetary scheme for the atom and used the notion of the quantum to explain its stability and its pattern of spectral lines. At face value, electrons should not have stable orbits around an atomic nucleus. Rather, because of their electrical attraction to the positive core, they should eventually spiral inward, emitting radiation as they plummet toward the center. The frequency of this radiation, if classical physics is any guide, should be synchronized with the frequency of the orbits.

However, that's not what happens at all. Atoms are relatively stable. Something must explain why electrons like to remain in steady orbits. Bohr brilliantly deduced that electrons' angular momenta must come only in discrete values—whole-number multiples of a constant known as $\hbar$, defined as Planck's constant divided by the quantity 2π. In other words, angular momentum, like energy, is quantized.

Angular momentum is a physical quantity that depends on an object's mass, velocity, and orbital radius. It comes into play when anything is spinning around—such as a ballet dancer pirouetting or a galaxy revolving. In classical physics it is a continuous parameter, meaning that it can take on any value. If a dancing coach tells a dancer to spin her partner a bit faster, she might pull a bit harder on his hand to give him the impetus (technically known as torque) to increase his angular momentum.

Remarkably, Bohr found that you can't make electrons revolve at just any rate or at any orbital radius. They can change states only by ingesting or expelling finite chunks of energy and angular momentum. Thus, rather than adjusting their locations or velocities continuously, the electron "dancers" suddenly switch from one position to another, like movement under a strobe light.

Changes in an electron's energy level happen whenever a photon is absorbed or emitted. The energy of a photon is its frequency times Planck's constant. That quantum of energy is gained or lost whenever a photon is acquired or expelled, respectively. Surprisingly, as Bohr demonstrated, the frequency of radiated photons has nothing whatsoever to do with the electron's orbital frequency (how many times it circles per second). Rather, it is an independent quantity that depends only on the size of the energy level jump with which the photon is associated.

Bohr's hypothesis of quantized angular momentum and energy enabled, for the first time, accurate predictions of the orbital radii and energy levels of electrons in a hydrogen atom. It offered a set of "Kepler's laws" (rules for planetary motion) for the atomic "solar system."

Although it certainly was incomplete—it addressed only hydrogen atoms and didn't justify its assumptions of quantized angular momentum and energy—it fit existing data well. Its major litmus test, which it passed with flying colors, was matching the Rydberg formula for the wavelengths of hydrogen's spectral lines.

Proposed in 1888 by Swedish physicist Johannes Rydberg, the Rydberg formula is a simple algorithm for deducing the patterns of wavelengths in atomic spectra. It predicts several distinct sequences of lines in hydrogen's spectrum, known as the Lyman series, the Balmer series, the Paschen series, and so forth. Bohr showed that these series—and the Rydberg formula in general—derived precisely from his set of assumptions about electrons and photons in hydrogen atoms. Each spectral line's wavelength matched the predicted value for a photon given off during an electron's transition between two different energy levels.

Bohr's model is now called the "old quantum theory." His ad hoc assumptions advanced our understanding of atoms but could not be explained through any known physical principles. It would take the careful work of Schrödinger, Louis de Broglie, Werner Heisenberg, and others in the 1920s to place quantum theory on much firmer footing.

Sketches of a Revolution

The most anticipated talk of the 1913 Vienna conference was a lecture by Einstein on the morning of Tuesday, September 23, entitled "The Present Status of the Problem of Gravitation." The great lecture hall was packed with those eager to hear the man who had published so many remarkable papers in a single year speak about his new theories. Einstein did not disappoint the crowd. He delivered one of his most important scientific addresses: a sketch of his ideas about a new explanation of gravitation that would transcend Newton's laws. Offering tantalizing morsels of Machian philosophy, bite-sized portions of higher mathematics, and an enticing prediction about starlight during solar eclipses, he gave the hungry audience a taste of what would become his exquisite general theory of relativity.

Einstein began his talk with a brief history of electromagnetism, starting with Coulomb's inverse square law of the electric force between charges. He showed how in the nineteenth century the contributions of

Faraday and others demonstrated deep connections between electricity and magnetism, culminating in Maxwell's equations. These ties formed a kind of unified field theory, Einstein emphasized, uniting two natural phenomena, once thought to be unrelated, in a single theory. Maxwell's equations, he pointed out, set a maximum rate of communication—the vacuum speed of light. To accommodate classical ideas about relative velocity to the new reality of an invariant light speed, the special theory of relativity was developed.

Now the time was ripe, Einstein continued, to tackle nature's other fundamental force, gravitation. At that point, gravitation was just at the stage that electricity had been at when Coulomb's law was proposed. Newton's inverse square law of gravitation, with its notion of "action at a distance," was analogous to Coulomb's idea and similarly incomplete. It was time, Einstein emphasized, to develop a full field theory of all of the natural forces, including gravitation, which did not involve the archaic idea that interactions happened instantaneously across vast distances. Special relativity mandated that gravitation could not be transmitted instantly between two remote massive bodies. The interaction certainly could not happen faster than the speed of light. Gravitation needed to be reframed, therefore, as a local field theory that respected nature's upper speed limit.

By making an analogy between electromagnetism and gravitation, Einstein was clearly sowing the seeds for a unified explanation of both of them. He wanted to continue Maxwell's program of melding different forces together by blending gravitation into the mix. Explaining gravity by itself would only be the first step.

Schrödinger listened intently to the words of the man who would come to be his mentor. Einstein's crystal-clear explanation of the deep connections between the forces opened his eyes to the stunning possibilities of fundamental theoretical physics. Subsequently, Schrödinger would see no limit to the kinds of problems he could tackle, including cosmic questions much broader than the atmospheric radiation measurements on which he had been focused.

At that time, Schrödinger was one of the few scientists who appreciated the full scope of Einstein's ambition to unify the natural forces. "The conception Einstein put forward," he would later write, "embraced from the outset (and not only by the numerous subsequent attempts to generalize it) every kind of dynamical interaction, not just gravitation only."[4]

In tandem with his broader ambitions in physics, Schrödinger would soon begin to read much philosophy. His eyes became focused on signs of unity in nature. An ensuing search for unifying principles would lead him to the work of the nineteenth-century German philosopher Arthur Schopenhauer, Eastern mystics, and others who sought to explain the underlying mechanisms of the universe.

Certainly Schrödinger could well relate to Einstein's interest in the philosophy of Mach, who was sickly and long retired but still actively interested in science. Einstein molded Mach's critique of the Newtonian concept of inertial frameworks (constant speeds relative to "absolute space") and his vague alternative idea that distant pulls from the stars cause inertia into a specific connection between mass and inertia. In Einstein's interpretation of Mach, the collective mass of all bodies in the universe exerts an influence on objects, causing them to move naturally in straight lines at constant speeds. Hence inertia is an aggregate effect of the distribution of mass in the universe—something like the dim nocturnal haze that is the combined influence of streetlamps in a city. (When the conference wasn't in session Einstein would visit Mach in his Vienna apartment and discuss mutual scientific interests with the elderly, gray-bearded philosopher.)

In the most technical part of his talk, Einstein proceeded to sketch his ideas, developed in mathematical form with Grossmann, for a way of connecting the mass distribution in space to its four-dimensional geometry, and finally to the local motions of objects in the form of what we call gravitational acceleration. He pointed out that his theory rests on the idea that inertial mass (how an object accelerates in response to forces) is precisely equal to gravitational mass (how an object is attracted to others through gravity). This leads to a cancellation in the equations of motion of an object's own mass, meaning that at a particular point of space any massive object would display the same behavior. Consequently, an object's location—and the geometry of space at that point as shaped by the mass distribution in the universe—governs its behavior.

The culmination of Einstein's talk was a bold, testable prediction about the bending of starlight by the Sun. He forecast that the Sun's massive presence would warp the geometry around it, causing everything in its vicinity to move in curved paths (from our perspective) rather than straight lines. Even the light emitted by distant stars would bend as it neared the Sun. Tracing these rays backward, we would perceive these stars to shift to different positions than they would appear at if the Sun's bulk wasn't there. Naturally, because we normally can't observe the stars by daylight, we wouldn't ordinarily witness this gravitational light-bending effect. However, during a total solar eclipse, Einstein pointed out, the "relocation" of stars would be eminently observable. He suggested that this distortion be measured, and matched against his theory, during an impending eclipse due to take place over Eastern Europe in August 1914.

The Vienna conference deeply impacted Schrödinger's career. Veering away from experimental measurements of radiation, he began to move in a theoretical direction, examining fundamental questions about physics. However, before he could delve deeply into research in atomic physics, gravity, and the other topics he was exposed to during the conference, fate would intervene.

On June 28, 1914, Archduke Franz Ferdinand, designated heir to the Austro-Hungarian crown, was visiting Sarajevo, Bosnia, when Serbian nationalist Gavrilo Princip fatally shot him and his wife. One month later, the First World War began and Schrödinger received his marching orders. He served loyally on the Italian front, performing various duties such as commanding a battery. While Germany joined the war, fighting on the side of Austria-Hungary, Einstein adamantly opposed the conflict and refused to participate.

Returning to Vienna in spring 1917, Schrödinger continued his military obligations by doing meteorological work alongside his friend Hans Thirring. Sadly, the war had delayed Schrödinger's academic career by about a third of a decade—a frustratingly long time for a young researcher. Back in Vienna, he was eager to make up the lost time by resuming his theoretical endeavors and teaching.

The war would also delay the testing of Einstein's light-bending prediction. German astrophysicist Erwin Finlay-Freundlich, a student of Klein and ardent follower of Einstein's theories, enthusiastically mounted an expedition to the Crimea, where the solar eclipse would be prominent, with hope of recording the phenomenon. But before he could take the measurement, the Russian army captured and interned him as a prisoner of war. It would take another half decade, after the war was over, for such a test to be performed and Einstein's hypothesis confirmed. In the interim, Einstein would continue to develop his theory of gravitation.

The Happiest Thought

The roots of the general theory of relativity date back well before the 1913 conference. In 1907, only two years after his special theory was published, Einstein had what he later called the "happiest thought of [his] life." As he recalled: "I was sitting on a chair in my patent office in Bern. Suddenly a thought struck me: If a man falls freely, he would not feel his weight. I was taken aback. The simple thought experiment made a deep impression on me. It was what led me to the theory of gravity."[5]

Einstein had stumbled upon the principle of equivalence: a simple but powerful idea that became the bedrock of general relativity. It derives from the idea that because inertial mass equals gravitational mass, all objects accelerate equally under pure gravity. Legend has it that Galileo dropped a stone and a feather from the Leaning Tower of Pisa to test whether this was true. Because it is, free-falling objects, accelerating downward at precisely the gravitational acceleration, appear to be weightless. That's because if an object were falling along with a scale directly under it, both would journey downward at the same rate, so the scale would not feel a weight on it. Roller coaster enthusiasts note such a sensation of weightlessness during downward plunges.

Einstein took this idea a step further by stating that no physical experiment could distinguish a free-falling body from one that is at rest (assuming that no other forces such as air resistance interfere). Therefore, if a girl plummeting straight downward in a free-fall ride at an amusement park were playing the violin, shuffling cards, or stacking building blocks, she would find the task just as simple or challenging, as the case may be, as performing the same feat while at rest. That's because everything would be accelerating downward along with her at precisely the same pace.

Einstein brilliantly realized that it would be possible to patch together a comprehensive theory of gravity from free-falling reference frames, treating each as if it were at rest. Within each framework, he noted, each object would move in a straight line unless deflected by external forces. However, when viewed from another framework, those straight lines might appear curved. That is why we see bodies taking curved paths because of gravity—because we are viewing their motion from our framework, not theirs.

To understand how this works, let's revisit the "light ping-pong" analogy mentioned in Chapter One. Imagine that an astronaut is bouncing a beam of light off a mirror on one side of his transparent spaceship, while his sister is using a hypothetical ultrapowerful telescope to watch him from Earth. Suppose his ship is free-falling toward a planet. From his perspective, the light ray would travel across his ship in a perfectly straight line. If he transmitted the light horizontally across the ship from a height of three feet, it would hit the mirror at a height of three feet as well. However, from his sister's point of view the ship would be falling and the light would be bending downward. By the time the light reached the mirror, the ship and mirror would be much lower. Thus the light would take a curved path—from a high starting point to a much lower bounce off the mirror.

This phenomenon led Einstein to make his prediction about starlight curving near the Sun during an eclipse, even before he had the mathematical skills to beef up his theory with a solid geometric framework. At first he tried more moderate tweaks to the special theory, including the idea of making the speed of light vary from point to point. However, he couldn't get the math to work as he wished. He started to think about more sophisticated mathematical methods, such as changing the metric—the components of the formula for calculating distances—but didn't yet have the knowledge to complete his scheme.

Sometime in late 1912, Einstein became aware of a body of experimental results published by Hungarian physicist Baron Loránd (Roland) von Eötvös about the equivalence of inertial and gravitational mass. Einstein had suggested such an experiment himself before learning about Eötvös's extensive studies. Over the course of many decades, Eötvös had perfected an instrument called a torsion balance, designed to pick up even subtle differences between mass's inertial and gravitational values. In various versions of the experiment, he reached higher and higher precision and still found absolutely no discrepancy. For Einstein, Eötvös's work showed that the principle inspired by his "happiest thought" was not just an abstraction but rather a deep, empirical truth about nature. The "Old One"—as Einstein often personified his notion of an equation-developing deity—had left a vital clue, and it was his job to solve gravity's sphinxlike riddle.

Pulled Out of the Quagmire

In July 1912, after working for about a year at the University of Zurich and a little more than a year at the University of Prague, Einstein returned to the city of Zurich to begin a position at his alma mater, ETH. One of the main attractions, along with being in his beloved Switzerland, was working alongside his friend Grossmann, who was a math professor. The new position turned out to be fortuitous for the development of general relativity. Einstein was sinking quickly into the quicksand of confusion about higher mathematics and needed a strong grip to pull him back to safety. The same former classmate who had assisted him with his math in college proved indispensable in his quest to understand gravity geometrically.

Grossmann had little interest in physics but became passionate about Einstein's project. He gave Einstein a crash course in the works of Riemann, including how to manipulate tensors that describe the properties of non-Euclidean, higher-dimensional manifolds. (Recall that tensors are mathematical objects that transform in a certain way and that manifolds are surfaces that can have any number of dimensions.) He also introduced Einstein to the papers of German mathematician Elwin B. Christoffel, Italian mathematician Gregorio Ricci-Curbastro, and Ricci-Curbastro's student Tullio Levi-Civita, each of whom had contributed to the differential calculus of curved geometries.

Grossmann's extensive assistance gave Einstein new optimism about overcoming his difficulties in expressing his ideas mathematically. Einstein worked on the theory with a fevered intensity, temporarily abandoning all of his other scientific interests. When Sommerfeld invited him to Munich to give a talk about quantum theory, he declined, writing back: "I am now occupied exclusively with the gravitational problem, and believe that I can overcome all difficulties with the help of a local mathematician friend. But one thing is certain, never before in my life have I troubled myself over anything so much, and that I have gained great respect for mathematics, whose more subtle parts I considered until now, in my ignorance, as pure luxury! Compared with this problem, the original theory of relativity is childish."[6]

At one point Einstein came to Grossmann's apartment so often in the evening, the latter's elderly maid grew weary of running down the stairs to open the front door. Einstein's solution was to ask Grossmann to "leave the front door open so that the old girl won't have to bother."[7]

Within a year, Einstein and Grossmann had worked out the preliminary version of their theory that Einstein would present at the 1913 Vienna conference. Historians refer to this early form as the *Entwurf* or "outline," from the name of a paper they published around that time, "Outline of a Generalized Theory of Relativity and Theory of Gravitation." It contained many, but not all, of the elements of what became general relativity.

In special relativity, observers traveling at constant velocities relative to each other experience identical laws of physics. For example, Maxwell's equations appear the same for both. One of Einstein's key goals in formulating general relativity was to extend the concept of universal laws to observers accelerating relative to each other. Unlike Newtonian mechanics, which favors inertial or nonaccelerating frameworks, Einstein wanted his theory to be one-size-fits-all. A researcher in a lab that happens to be on a train grinding to a halt or on a carousel spinning in a circle should be able to describe her experiment by means of the same physics as one who is working in an ordinary stationary brick-and-mortar facility. Mathematically, that means that the equations should have the same form for accelerating coordinate systems—including speeding up, slowing down, or rotation—as for nonaccelerating coordinate systems. Einstein called this condition "general covariance."

Unfortunately, Einstein came to realize that the *Entwurf* didn't meet his objective of being independent of the coordinate system. It did not live up to his Machian goal of abolishing the preference for inertial frameworks altogether and establishing a kind of democracy for all kinds of motion, including acceleration. Rather, there was still an elite in which certain kinds of coordinate systems were favored.

Einstein turned to another former classmate, Michele Besso, for advice on the scientific validity of the *Entwurf*. If the theory was physically correct, perhaps he could live with certain mathematical limitations, such as a lack of general covariance. Einstein was doggedly persistent about his ideas but would abandon them in a flash if he could see a more economical way forward. He tried to persuade himself, for a time, that general covariance was not necessary for a complete theory, as long as the equations were simple and produced physically valid results.

Besso and Einstein decided to see how the *Entwurf* treated a benchmark of astronomy: the rate of precession (advancement in the direction of rotation) of Mercury's perihelion (closest point to the Sun). As the innermost planet, Mercury feels the Sun's gravity the strongest and is therefore the most sensitive to tests of gravitational theories. While Newton's gravitational theory describes very well the motions of the other planets in the solar system, it fails to account for the Spirograph-like turning of Mercury's elliptical orbit, advancing slowly over the eons, retracing the same pattern once every 3 million years. Einstein hoped that the *Entwurf* would produce a more accurate prediction. To his dismay, calculations performed by Besso indicated that his theory still generated an inaccurate precession rate.

Another prediction Einstein and Grossmann made in the *Entwurf,* the deviation of starlight because of the Sun's massive influence, would have been tested by Finlay-Freundlich during the 1914 solar eclipse if he hadn't been captured. If he had been able to take that measurement, he likely would have found that the *Entwurf* offered inaccurate results for that value too. The theory clearly needed an overhaul, forcing Einstein to struggle with the equations much longer than he had anticipated.

The struggle would need to proceed without Grossmann by his side and also without Mileva. Because of Albert's focus on his work, to the exclusion of his family life, and Mileva's plunge into a deep depression, their marriage had been crumbling for some time. A career move back to Germany proved the final straw. He received an offer from Max Planck and physicist Walther Nernst to take on three important positions in Berlin—member of the prestigious Prussian Academy of Sciences, professor at the University of Berlin, and director of a newly established physics institute. One of the perks would be that he would no longer have to lecture; he could research his theories to his heart's content. After reluctantly following Albert to Berlin in April 1914, Mileva stayed there miserably for a few months before deciding to move back to Zurich with the children. After some time, they began to negotiate for a divorce, a process that would take years.

Albert, meanwhile, had a new flame: his first cousin Elsa Lowenthal, to whom he would eventually get married. She was much more domestic and nurturing than Mileva. She often treated him like a child who needed to be fed, groomed, and generally cared for—for example, taming his unruly hair. She also proudly took care of maintaining his social calendar, relishing any opportunity to show him off in public. In turn, he was relieved to have his basic necessities attended to without fuss or argument so that he could focus on his calculations. He abided no interruptions in his tireless efforts toward a theory of gravitation, save the sweet strains of violin during his music breaks.

The Race to the Summit

Just before Einstein gaspingly reached the summit of his aspirations, he sensed that David Hilbert was racing to climb the same peak. In June 1915, Einstein spoke to an eager audience in Göttingen, including Hilbert, about his progress toward the general theory of relativity and the obstacles that remained, including the general covariance issue. Intrigued by the challenge of describing a non-Euclidean spacetime shaped by its matter and energy, Hilbert decided to pursue finding the field equations of general relativity himself. Suddenly Einstein faced the heat of competition. He was mortified that one of the ablest mathematicians in the world was eyeing the trophy he had been seeking for years. The race was close, but Einstein planted his flag first. By late autumn, he had joyously arrived at the correct formulation.

As a kind of consolation prize, however, Hilbert is recognized as being the author of an alternative way of framing general relativity, called the Lagrangian formulation. Mathematically, a Lagrangian is the difference between the kinetic energy (energy of motion) and the potential energy (energy of position) of a mechanical system, written in the form of a function of its coordinates.

One can picture the distinction between potential and kinetic energy by thinking about a spring gun. Push the spring back and the potential energy increases, indicating that it has more potential to fire. Release the spring and the kinetic energy increases, meaning it actually fires. The potential energy of position converts into the kinetic energy of motion. Now subtract the potential energy, expressed in terms of the position variable, from the kinetic energy, expressed in terms of the velocity variable, and you have a Lagrangian.

As the brilliant nineteenth-century Irish mathematician and astronomer William Rowan Hamilton showed, one can integrate (sum up through calculus) the Lagrangian over time to form a quantity called the action. Then, as Hamilton proved, any mechanical system evolves in such a way to minimize the action (or in some cases to maximize the action). This idea, called the least-action principle, naturally leads to the equations of motion, called the Euler-Lagrange equations. Therefore, in short, by knowing the Lagrangian of a system you can determine how it develops.

In classical mechanics a simple example of this is an object—such as a box of Tang discarded by astronauts decades ago—moving slowly through empty space without any forces on it. Its kinetic energy is just half its mass multiplied by its velocity squared. Its potential energy is zero, due to the lack of forces and the uniformity of empty space. Therefore, its Lagrangian is just its kinetic energy. The least-action principle shows that the object's path of minimal action is simply a straight line. Plug the Lagrangian into the Euler-Lagrange equations and the result is an equation stating that the velocity is constant. Therefore, the Tang box's rather simple Lagrangian dooms it to travel at constant speed in a straight line indefinitely.

Hilbert's contribution, called the Einstein-Hilbert Lagrangian (leading to the Einstein-Hilbert action), is fairly straightforward too. However, it is rich enough mathematically to generate Einstein's field equations of general relativity Moreover, if you have the inclination to alter general relativity in a physically reasonable way, tweaking the Lagrangian offers a means of doing so. We'll see that Schrödinger, in his efforts toward extending general relativity to encompass other forces, would eventually do just that.

Hamilton developed another way of describing mechanical systems, called the Hamiltonian method. Instead of subtracting the potential energy from the kinetic energy, the two quantities are added. This sum, called the Hamiltonian, can then be used with a set of equations to explore how a system's position and momenta relate to each other. Like the Lagrangian method, the Hamiltonian method has also come to play a vital role in modern physics, including, as we'll see, Schrödinger's formulation of quantum mechanics. The Hamiltonian tool kit can similarly be applied to general relativity, as Einstein showed when he finally wrapped up his theory.

A Glorious Edifice

Einstein debuted his masterpiece in nearly its final form at a meeting of the Prussian Academy on November 4, 1915. He was delighted to present the field equations for a comprehensive theory of gravitation based on spacetime geometry. On November 18, he gave another talk to the same group in which he offered his solution to the age-old problem of Mercury's orbital precession. Two month later, after calculations had conclusively verified his theory, he wrote to his friend Paul Ehrenfest, "Can you imagine my joy at the feasibility of the general covariance with the result that the equations of the perihelion movement of Mercury prove correct? I was speechless for several days with excitement."[8]

By the time Einstein published the final version of his theory in the prestigious *Annalen der Physik* on March 20, 1916, German physicist Karl Schwarzschild, while serving as a soldier on the Russian front, had already come up with the first exact solution. Remarkably, he had read a report of the November 18 talk, and computed the case of the gravity of a massive, spherical object such as a star. Amidst the darkness of war, Einstein's gleaming creation lit up the sky even brighter than a mortar, offering hope and inspiration to at least one soldier. Sadly, Schwarzschild developed a fatal autoimmune disease and died on May 11, 1916, at the age of forty-two. Many decades later the Schwarzschild solution would be used to describe black holes. Since then, numerous other exact solutions of Einstein's equations of general relativity have been found.

Einstein's golden temple is built on a gritty foundation: the material and energy content of the universe. Start with any matter and energy distribution, expressed in the form of what is called the stress-energy tensor, $T_{\mu\nu}$, and the field equations of general relativity tell you the components of another mathematical entity, representing the geometry of spacetime, called the Einstein tensor, $G_{\mu\nu}$. The equation $G_{\mu\nu} = 8\pi T_{\mu\nu}$ (which can be written in various ways) is considered one of Einstein's most important contributions, along with $E = mc^2$ and the equation for the photoelectric effect. All three equations are carved into the Einstein Memorial in Washington, D.C., as testimony to his brilliance.

An anecdote once told by acclaimed physicist Richard Feynman illustrates the ubiquity of Einstein's field equations in modern discussions of gravitation. Feynman was invited to the first American conference on general relativity, in Chapel Hill, North Carolina, in 1957. When he arrived at the airport and was about to take a taxi to the conference, he didn't know whether it was being held at the University of North

Carolina or North Carolina State. So he asked the taxi dispatcher if he had noticed any people looking distracted and muttering "*G mu nu, G mu nu.*"[9]

The gist of the Einstein equations is that the geometry of a region, expressed by the Einstein tensor, is determined by its matter and energy content, expressed by the stress-energy tensor. In other words, mass and energy warp spacetime, telling it where and how to curve. The shape of spacetime, in turn, governs how things move within it. Hence Einstein's equations beautifully connect the stuff of the universe with the shape of the universe.

Any tensor can be written in terms of its components in the form of a matrix, or array—something like a chessboard. The Einstein tensor and stress-energy tensor each can be expressed as 4-by-4 matrices. These have sixteen components each, but not all the components are independent. That is because a symmetry rule requires that if a component with a certain row and column number (the third row and fourth column, for instance) has a particular value, the component with those row and column numbers switched must be identical (the fourth row and third column). It is like arranging the pieces of a game of chess so they seem like they are reflected in a mirror along the diagonal of the board. We call such tensors symmetric.

With the symmetry rule in place, the Einstein tensor has ten independent components. So does the stress-energy tensor. Therefore, Einstein's equations that associate the two tensors lead to ten independent relationships between components. These show how matter and energy affect different aspects of space and time. Some of the relationships might lead to stretching or compression. Others might lead to twisting or turning. Anything that might happen to space and time, due to the gravitational effects of matter and energy, is there in the equations.

If Einstein's equations are so simple and elegant, why did it take him so long to develop them? As the saying goes, the devil is in the details. You can't just take the Einstein tensor and directly plot out the motions of astronomical objects such as planets or stars. The way things move is determined by yet another mathematical entity, this one called the metric tensor. To proceed from the Einstein tensor to the metric tensor is not at all obvious and requires several distinct steps.

Suppose you know the mass-energy distribution in a region and wish to determine how objects move through it. Here are the steps involved. First use Einstein's equations to find the Einstein tensor from the stress-energy tensor. Both the Einstein tensor and the related Riemann curvature tensor (the former is a kind of shorthand for the latter) encode information about the curvature of spacetime from point to point. Then use either the Einstein or Riemann tensor components to construct geometric objects called affine connections (also known as Christoffel connections). These dictate how the components of vectors (objects with magnitude and direction) transform as you move them— as parallel to themselves as possible—from point to point. Next, use the affine connections to reveal the components of the metric tensor. The metric tensor sews together the fabric of spacetime by specifying how to measure the distances between points. It offers a variation of the Pythagorean theorem for curved spacetime. Finally, use the metric to determine the most direct paths objects can take through space. Because of the warping of spacetime, these are generally curved, such as planets' elliptical orbits around the Sun.

Although the mathematics of general relativity can challenge even PhD students, let's use an analogy to illustrate its different layers. Start with a flat, boundless desert, representing empty spacetime. Upon the desert sand we scatter rocks of various sizes and weights, symbolizing the variety of massive objects in the universe, such as stars and planets. We find that the heavier rocks press down upon the sand far more than the lighter rocks, creating much deeper indentations. The areas without rocks remain flat. Hence, the more mass in a particular region, as recorded by the stress-energy tensor, the more it dips, representing greater curvature as measured by the Einstein tensor.

Now imagine that in our analogy you can't walk on the sand or rocks—they are too hot. Therefore, we need to construct a sturdy canopy above it, supported by a structure that respects the topography. We gather numerous poles (local coordinate axes) and bars (affine connections) to create the skeletal structure. The bars connect different poles in a way that guides how they are oriented. Similarly, affine connections determine how coordinate axes vary throughout space, in a configuration that depends on the rising or dipping of the base.

Finally, we weave a firm canopy designed to fit snugly upon the structure. In some places we need to sew neighboring points tighter together to make the fabric bend in certain ways. In other places, neighboring points are more loosely connected. The sewing pattern that determines how to stitch together the canopy in just the right way to match the skeletal structure just below it (and the rises and dips in the sand beneath that) symbolizes the metric tensor. Hence, we see how the metric tensor stitches the fabric of spacetime in a manner governed by the affine connections, which in turn depend on the Einstein tensor that is shaped by the stress-energy tensor. Got that?

Let's now take a walk on spacetime's canopy. We strive to take the fastest route possible, so we aim for a straight line. However, the canopy dips where there is a massive group of rocks underneath it, making even the most direct lines veer in various directions. Consequently, we follow a curved path, ringing around the indented area, tracing out an oval shape. Curiously, we have gone into orbit—like young Schrödinger around his aunt when they played their planet game.

The Forever Universe

Once his general theory of relativity was complete, Einstein decided to apply it to the universe in its entirety. His goal was to show that the universe is a relatively stable collection of stars and other bodies. True, the stars move, he realized, but they do so slowly. Einstein's proposed cosmology would offer the permanence and steady framework of Newtonian "absolute space" without resorting to what he—following Mach—viewed as fictitious.

Einstein decided to begin his cosmological calculation with the basic assumption that space is isotropic, meaning uniform in all directions. He chose a simple four-dimensional geometry called a hypersphere to represent the shape of space. A hypersphere is a generalization of a sphere to an extra dimension. If you live in a hypersphere and journey in any direction, you'll eventually return to your starting point—just as if you circled the equator on Earth. The advantage of the universe having a hyperspherical profile is that it would be finite but free of boundaries. Only someone standing beyond the universe would notice its "surface." Within space, there would be no limits, just repetition. The Argentine writer Jorge Luis Borges would wonderfully represent this concept in his imaginative short story "The Library of Babel," in which he envisioned the cosmos as a vast but finite and repetitive collection of books.

Einstein attempted to find a static solution to his field equations but soon realized that there was a problem. The only solution he found was unstable. If nudged a bit, by changing the distribution of matter slightly, it would either collapse or expand—like a balloon being popped or inflated. To replicate an eternal, stable universe, such a solution certainly wouldn't do. Edwin Hubble's discovery of cosmological expansion—what we now call the "Big Bang"—would be more than a decade in the future. Therefore, Einstein reasonably believed that space was static and saw expanding models as unphysical.

To remedy the situation, he took the rather drastic step of adding an extra term to the geometric side of his equations to produce what he saw as credible solutions. Known as the "cosmological constant" and symbolized by the Greek letter lambda (Λ), the term acts as a hedge against gravitational instability by stretching the geometry of space in the opposite direction. He didn't assign the cosmological constant any physical meaning, but at the time he saw it as essential to the integrity of his theory.

In our desert canopy analogy, imagine that the entire structure we had constructed was slowly sinking into the sand. Rather than rebuild the structure from scratch, we might elect to build mechanical devices around the periphery that grab the canvas and stretch it outward. We wouldn't win any architecture awards for our design, but it would do the job. Similarly, the cosmological constant term, through inelegant, performed the task Einstein set out to accomplish: preserving cosmic stability.

In 1917, Einstein published his static universe model, including the cosmological constant as part of his field equations. However, he couldn't rightfully claim that his solution was unique. Dutch mathematician Willem de Sitter cleverly demonstrated that in the absence of matter, Einstein's field equations produced solutions that would blow up exponentially, driven ever outward by the cosmological constant. De Sitter's model showed that as long as the cosmological constant exists, emptiness is unstable. Given that he added the cosmological constant term as a stopgap, rather than based on scientific observation, Einstein did not take de Sitter's model very seriously. He conceded, however, that progress in understanding the dynamics of the universe would require far more astronomical measurement. Luckily, Hubble would do just that, with his giant reflecting telescope on Mount Wilson in southern California, ultimately revealing a cosmos that was expanding rather than static.

Anticipations of Dark Energy

One might argue that Ernst Mach, who died in 1916, would have dismissed the idea of adding a term to the equations of general relativity that had nothing to do with sensory experiences. Just as Newton introduced absolute space just to define inertia, Einstein's inclusion of a cosmological term was decidedly un-Machian. It took another follower of Mach—Schrödinger—to suggest a more tangible alternative.

Schrödinger first became aware of Einstein's complete field equations of general relativity in late 1916, when he was commanding a battery at Prosecco during the war.[10] By the time he returned to Vienna in spring 1917, he found that many of his colleagues at the university, including Thirring, were busy trying to find ways of interpreting and applying Einstein's theory. For example, along with Austrian physicist Josef Lense, Thirring showed how rotating objects affect the spacetime around them—a result known as "frame-dragging" or the Lense-Thirring effect.

In November 1917, Schrödinger submitted two papers to the German journal *Physikalische Zeitschrift* exploring different aspects of general relativity. The first dealt with the question of defining gravitational energy and momentum from point to point in a way that is independent of the choice of coordinate system. He examined Schwarzschild's solution and showed that one way of framing gravitational energy produced the surprising result that the object had no energy at all. Interestingly, the issue Schrödinger brought up anticipated decades of debate about the question of defining energy consistently in general relativity.

Schrödinger's second work, "Concerning a System of Solutions to the Generally Covariant Equations for Gravitation," addressed head-on the issue of the physicality of the cosmological constant. He questioned the placement of an extra term on the geometric (Einstein tensor) side of Einstein's equations, arguing that the same effect could be achieved through a modification of the material (stress-energy tensor) side instead. As Schrödinger noted, "The completely analogous system of solutions exists in its original form without the terms added by Herr Einstein. The difference is superficial and slight: The potentials remain unchanged, only the energy tensor of matter takes a different form."[11]

The extra "tension" (stretching) term Schrödinger posited served to counteract the gravitational effect of matter by adding a kind of negative energy and making the mass density effectively zero. With a zero mass density throughout space, the universe would no longer be compelled to experience gravitational collapse and would thereby maintain its stability. He justified the zero mass using a Machian argument that mass is noticeable only when it is excessive. The argument is akin to saying that we generally notice black and white hues only when they are contrasted with other colors. We might refer to a perfectly black or white sky as having no color at all.

Einstein soon published a reply to Schrödinger's cosmology paper—the beginning of a scientific dialogue that would take many twists and turns over multiple decades. He pointed out that Schrödinger's hypothesis allowed for two options: a new constant term or a new type of energy with negative density that varied from point to point. The former case, Einstein argued, was equivalent to the cosmological constant term, only on the other side of the equations. The latter, on the other hand, would be unphysical (because it would have a negative energy density) and tricky to map out. As Einstein wrote: "One not only has to start out from the hypothesis of the existence of a nonobservable negative density in interstellar spaces but also has to postulate a hypothetical law about the space-time distribution of this mass density. The course taken by Herr Schrödinger does not appear possible to me because it leads too deeply into the thicket of hypotheses."[12]

Interestingly, the concept of a substance with negative energy density—or, alternatively, negative pressure—has emerged in recent years as a possible solution to a cosmological puzzle. In 1998, two teams of astronomers augmented Hubble's discovery with their own finding about cosmic growth. They discovered that not only is the universe expanding, but the rate of expansion is speeding up. Some unknown agent is causing the universe to accelerate. University of Chicago cosmologist Michael Turner dubbed that entity "dark energy."

Curiously, the suggestion made by Schrödinger and discussed by Einstein about a substance counteracting gravity would do the job just nicely. For that reason, science historian Alex Harvey has recently suggested that Einstein discovered the concept of dark energy.[13] "Discovered" may be too strong a word, given that there was no real physical motivation at that time. More precisely, in 1917 he envisioned that such a negative-energy substance was within the realm of possibility—never dreaming that the universe was actually accelerating due to some unknown cause. Nevertheless, it is interesting that the groundwork for such a notion was laid down so early.

World Famous

When the First World War ended on November 11, 1918, it left behind a Europe that was barely recognizable. Empires fell, borders changed, new leaders emerged, and conditions began to develop that would set the stage for another world war. The Austro-Hungarian Empire was replaced by a number of smaller states, including Austria (originally called "German Austria"), Hungary, and Czechoslovakia. A democratic but much weakened Weimar Republic controlled much of what had once been the German Empire. The victorious Allied Powers were determined that Germany should pay the price for the bloody war of attrition. It was forced to cede some of its territory, limit the size of its army, and pay extensive reparations—leading to much resentment and economic depression that would contribute to the rise of the Nazis.

During the war, Einstein had little chance to test his hypothesis about the gravitational bending of starlight by the Sun. Finlay-Freundlich's inability to complete his expedition was a great disappointment to him. Einstein quietly began to correspond with a British astronomer, Arthur Eddington, who was keenly interested in verifying Einstein's theory. According to several widely reported stories, Eddington was known at the time as one of the few people who truly understood general relativity.[14]

A Quaker and a pacifist, Eddington, like Einstein, was opposed to the war and in favor of international scientific cooperation. Naturally, during the bloody conflict, open cooperation between British and German scientists was close to impossible. The armistice opened up a grand opportunity for Eddington to help test Einstein's theory and thus rekindle the trust between scientists of their respective nations.

Eddington and Frank Watson Dyson, Astronomer Royal of Britain, realized that an ideal opportunity to measure gravitational light bending would arise on May 29, 1919. On that day, a solar eclipse would occur over part of the Southern Hemisphere just when the Sun was passing in front of the Hyades star cluster, a particularly bright formation. Dyson appointed Eddington to be the organizer of a project to observe the eclipse, a move that helped save the latter from internment as a conscientious objector.[15]

In January 1919, to set a baseline for the observation, Eddington carefully measured the unaltered positions of the Hyades stars. Then he organized two expeditions to record their sky positions during the eclipse. The first team, led by Eddington himself, traveled to Príncipe, an island in the Gulf of Guinea off the west coast of Africa. As a backup in case of inclement weather, a second group was sent to Sobral, Brazil. The two teams carefully photographed the new positions of the stars and brought their data back to Britain for a detailed comparison with the original results. Completing his analysis on November 6, Eddington was delighted to announce that the angular deviations, averaging 1.61 arc-seconds for Príncipe and 1.98 arc-seconds for Sobral, came close to Einstein's general relativistic prediction of 1.75 arc-seconds and were far greater than an estimate based on Newtonian theory of half that amount.

At a meeting of the Royal Society chaired by Dyson, a packed audience hailed the results, acknowledging them, along with the Mercury precession findings, as important proof of general relativity. In an age of political revolution, the eclipse results showed that science too had been swept up in colossal changes. For a group of British scientists to acknowledge one year after the war's end that a German physicist had overthrown Newton was truly extraordinary. As Thomson proclaimed: "These are not isolated results. . . . It is not the discovery of an outlying island, but of a whole continent of new scientific ideas of the greatest importance to some of the most fundamental questions connected with physics. It is the greatest discovery in connection with gravitation since Newton enunciated that principle."[16]

It is an emblem of how little Einstein was known internationally before the announcement that a *New York Times* article about the discovery referred to him simply as "Dr. Einstein, Professor of Physics in the University of Prague."[17] Not only did the article fail to mention his first name, it also got his affiliation wrong, as it had been more than seven years since he had stepped down from his Prague position.

In a flash, Einstein became world famous. By toppling Newton, he had become a celebrity in his own right. Fame in the twentieth century was a far bigger deal than it had been in Newton's time. News traveled much faster in the age of wireless than in the era of the hand-cranked printing press. Newspapers around the world echoed the message of the stunning *Times* of London's three-line banner headline: "Revolution in Science . . . New Theory of the Universe . . . Newtonian Ideas Overthrown."[18]

The Lofty Clouds of Pure Geometry

The paint was barely dry on Einstein's masterpiece when he began to perceive its imperfections. As he stared at his achievement, the two sides of his field equations seemed imbalanced. On the left-hand side was a delicate representation of gravity's geometric patterns. On the right side, all types of matter and energy, including the energetic effects of electromagnetic fields, were coarsely lumped together in the stress-energy tensor. Einstein had the utmost respect for Maxwell's equations of electromagnetism and didn't like seeing them have a secondary role. He came to believe that electromagnetic fields should be rendered through geometry much the way gravity was, instead of merely being included in the stress-energy tensor. Memories of the geometry primer of his youth and the love of geometry he had developed through his contacts with Grossmann and others motivated him to write all of nature's laws through geometric principles.

Extending the sequence of special and general relativity, Einstein believed that a third breakthrough would be needed to complete the transformation of natural law and unite electromagnetism with gravity. Maxwell's equations and gravitational theory would then be special cases of a complete unified field theory, constructed wholly through geometric relationships.

Schrödinger would come to agree with Einstein's view that general relativity was incomplete as long as electromagnetism was omitted from the geometry side. "We are in patent need of field laws for the . . . electromagnetic field," Schrödinger would write, "laws that one would also conceive as purely geometric restrictions on the structure of space-time. These laws the theory of 1915 does not yield, except in the simple case of purely gravitational interaction."[19]

As Einstein began to embrace pure geometry—instead of geometry guided by material effects—his interest in experimentation started to wane. While his general relativity papers and talks strongly emphasized the need for experimental verification—through the Mercury precession, the bending of light, and another effect called the gravitational redshift—his movement toward a unified field theory would see him changing his rhetoric to more abstract arguments. Ironically, the university student who had loved being in the lab and had skipped many of his classes in higher mathematics because they seemed irrelevant had grown into an advocate of using mathematical beauty and sheer reasoning to guide his theories. As he would state in a lecture, "On the Method of Theoretical Physics": "Experience remains, of course, the sole criterion of the physical utility of a mathematical construction. But the creative principle resides in mathematics. In a certain sense, therefore, I hold it true that pure thought can grasp reality, as the ancients dreamed."[20]

Researchers associated with the Göttingen school of thought, which emphasized pure geometric reasoning, helped shape Einstein's growing interest in more abstract mathematical constructs. For example, Ehrenfest, a friend and confidant who became so close to Einstein that they were like brothers, was a key influence. Ehrenfest had studied in Göttingen and taken courses with Klein. He and his mathematician wife, Tatyana, whom he had met in one of Klein's classes, were greatly interested in the relationship between geometry and physics. They offered their home in Leiden, Holland, as a refuge where Einstein could escape Berlin, ponder theoretical dilemmas, and relax by playing chamber music (Einstein on violin, Ehrenfest on piano). Adept at asking poignant questions that revealed the essence of problems, Ehrenfest lent a ready ear as Einstein struggled to incorporate electromagnetism into general relativity.

Klein himself, though retired, took an interest in the treatment of gravitational energy and momentum in general relativity. Like Schrödinger in his first paper of November 1917, Klein thought that these quantities needed to be defined in a way that didn't depend on coordinate systems. All observers, he argued, should measure the same values of gravitational energy and momentum. Klein corresponded with Einstein in 1918 about the question. While Einstein wouldn't budge on his definition, Klein's comments likely further motivated him to place gravity and electromagnetism on equal footing. Having different definitions of energy and momentum for the two forces was a quick fix and by no means a satisfactory long-term solution.

Hilbert, Klein's prize appointment and arguably the greatest systematizer of geometry since Euclid, certainly had a profound influence on Einstein.[21] Einstein took note that Hilbert's formulation of general relativity attempted to unify gravitational and electromagnetism along the lines of a suggestion made by German physicist Gustav Mie that the electron is a kind of stable bubble within the electromagnetic field. Following Mie, Hilbert argued that matter didn't exist independently, but rather was the result of clumpiness within energy fields. These fields, in turn, could be described geometrically. Einstein didn't accept Hilbert's arguments at first but gradually came to believe that geometry was more fundamental than matter.

Looking at electrons and other matter particles as the products of geometry is akin to explaining knots in ropes by understanding how they tangle. Imagine a girl finding a curiously shaped knot in a ball of yarn and thinking it is something separate from the yarn. She asks her mom for a box of knots to play with. Her mother, who happens to be a professor at Göttingen, patiently points out that knots aren't separate things and shows her how the yarn can be twisted up to make more of them. Yarn is fundamental; knots are not. Similarly, Hilbert and Mie envisioned a natural order in which the geometry of fields is foremost and twists manifest themselves as particles.

One of Hilbert's most gifted students was German mathematician Hermann Weyl (known to his friends as "Peter"), who received his doctorate from Göttingen in 1908. After receiving his *Habilitation* in 1913, Weyl was appointed to ETH in Zurich, where he and Einstein were briefly colleagues and came to know each other. In 1918 Weyl published an epic account of general relativity and its possibilities, entitled *Space, Time, Matter,* which he would update several times in revised editions as his ideas developed. He sent an early copy to Einstein, who called it a "symphonic masterpiece."[22]

Gratified by Einstein's praise for his book, Weyl hoped that a new paper he had written, "Gravitation and Electricity," would draw the same enthusiastic reaction. The article offered a means of tweaking general relativity to include Maxwell's equations as a consequence. He sent Einstein the manuscript, hoping it would be recommended for publication.

While at first Einstein was pleased that Weyl seemed to have found a way of sneaking electromagnetism into the theater of gravitation, he drew back when he saw how much the intrusion would disrupt the show. Weyl's idea involved modifying the idea of how vectors behave in a process called parallel transport (being moved parallel to themselves from point to point). In standard general relativity, the affine connections that show how vector components transform and the metric tensor that determines how spacetime intervals (four-dimensional distances) are measured depend on each other in a direct mathematical relationship. In our desert canopy analogy, that's a direct link between the scaffolding and the tarp. Weyl tampered with that linkage by adding an extra factor, which he called the "gauge." Just as railroads in different countries (such as Russia and Poland) have different gauges (in that meaning, the distance between the rails), Weyl envisioned changing the four-dimensional distance standard from point to point in space. The bonus was that the addition of the gauge factor produced an effect equivalent to the electromagnetic field. However, Einstein thought changing the distance standard was unphysical, and he couldn't sanction such a radical change to his theory. Weyl was extremely disappointed that Einstein rejected his idea.

Although it was never incorporated into general relativity, Weyl's gauge idea was later applied to a different domain, particle physics, where it found much success. In the modern concept, instead of actual space, the gauge factor pertains to a kind of abstract space. Contemporary interest in the Higgs boson—essential for explaining the rest mass of certain particles—owes much to Weyl's gauge concept.

Adventures in the Fifth Dimension

Yet another Göttingen graduate, Finnish physicist Gunnar Nordström, had proposed his own unified theory in 1914. It was remarkable because it was the first theory to include a fifth dimension, supplementing the three dimensions of space and the dimension of time. Nordström found that adding the extra dimension offered the extra space in the theory needed to include Maxwell's equations of electromagnetism along with gravity. The theory was not based on general relativity, however, leading Nordström himself to abandon the idea two years later after he acknowledged the superiority of Einstein's approach. While there is no indication that Einstein took heed of Nordström's notion for unification, another five-dimensional idea left an indelible impression on him.

In April 1919, Einstein received a letter from Theodor Kaluza, a little-known *Privatdozent* at the University of Königsberg. (In the

German academic system, a *Privatdozent* is an instructor who earns his living by offering lectures to which tickets are sold, rather than being paid by the university.) At that low rank, which he held for twenty years, Kaluza barely earned enough to support his family. Cognizant, perhaps, of the modest beginnings of his own career, Einstein gave his full attention to the note despite its sender's humble status.

Although he was at that time far afield from the mainstream community, Kaluza had once experienced the mind-bending atmosphere of Göttingen. He had spent a year there in 1908 during his student days, becoming thoroughly infused with the geometric visions of Klein, Hilbert, and Minkowski. He also met future fellow unifier Weyl.[23] The seed was planted in Kaluza's brain for a unique approach to unification that would sprout eleven years later.

Kaluza's letter outlined an idea that had come to him as a kind of revelation. One day, when he was sitting in his study, it had occurred to him that by adding an extra dimension and extra components to the tensors of general relativity, the machinery of Einstein's equations would crank out a version of Maxwell's equations in addition to the gravitational factors. Instead of being a 4-by-4 array, the Einstein tensor would become a 5-by-5 matrix. Rather than having sixteen components, with ten of those independent because of symmetry reasons, it would have twenty-five, of which fifteen would be independent. That means there would be five additional independent components, four of which could be used to describe electromagnetism; the fifth was basically ignored. A simple change in the number of dimensions seemed to allow enough room for unification. As his son (who was in the room with him at the time) recalled, Kaluza was so excited that he froze in place for a few seconds, then jumped up and started humming a tune from *The Marriage of Figaro*.[24]

Both Nordström's and Kaluza's schemes, derived independently, relied on the idea of extending spacetime by an extra dimension. For mathematicians or mathematical physicists accustomed to breathing the heady air of Göttingen, envisioning higher dimensions was as simple as counting. One dimension is for lines, two is for squares, and three is for cubes. Add one more spatial dimension and you get hypercubes. Just as a cube is a three-dimensional object bounded by six squares, a hypercube is a four-dimensional object bounded by eight cubes. Tack on the dimension of time, and you have a five-dimensional entity—with time typically marked as the fourth dimension and the extra spatial dimension as the fifth.

Yet for a mainstream experimental physicist of that era, the concept of a fifth dimension would have seemed like something from the pages of H. G. Wells or a pulp magazine rather than genuine science. Aside from time, there was no direct visual evidence for any dimensions beyond length, width, and height. A theory using five dimensions would have seemed like postulating a way of walking through walls or conjuring gold from thin air.

Kaluza anticipated the naysayers by placing a "cylinder condition" in his theory that would make direct observation of the fifth dimension impossible. Just like a hamster spinning a wheel beneath its feet and getting nowhere, in Kaluza's theory all observable quantities are fixed with respect to changes in the fifth dimension. As much as the fifth dimension cycles, there is no noticeable effect, except for its indirect action in bringing electromagnetism into general relativity. The whirling is safely behind the scenes, forestalling any objections by experimentalists.

Einstein's first reaction was to praise Kaluza's paper as being far superior to Weyl's. Unlike Weyl's theory, it didn't seem to tamper with known facts about the universe such as the magnitudes of spacetime intervals. However, after he performed some calculations based on Kaluza's theory, Einstein's enthusiasm faded. In trying to describe how electrons move under the combined influence of electromagnetism and gravity, he couldn't find a reasonable solution. Instead, he crashed into the mathematical barrier known as a singularity—a place where one or more quantities blow up and become infinite. Somehow, like an aching tooth, the sore point needed to be extracted.

By pointing out the shortcoming of Kaluza's theory, Einstein highlighted a new impetus for trying to extend general relativity: developing a theory of how electrons move through space. Bohr's atomic model matched the main spectral lines of simple elements such as hydrogen by showing how quantizing angular momentum and energy restricts electrons to particular orbits. However, it did not offer a complete theory of how electrons behave otherwise—for example, when streaking through cathode ray tubes. As he began to explore the ramifications of various unified approaches, Einstein used the electron dilemma as a touchstone.

Eddington fully agreed with Einstein on the importance of the electron problem. Using Weyl's theory as a starting point, Eddington proposed an alternative unified theory based on altering the affine connection and establishing a four-dimensional geometry distinct from Riemann's. Still, he wasn't sure if his theory adequately explained electron motion. As Eddington wrote, "In passing beyond Euclidean geometry, gravitation makes its appearance; in passing beyond Riemannian geometry, electromagnetic force appears; what remains to be gained by further generalisation? Clearly, the non-Maxwellian binding forces which hold together an electron. But the problem of the electron must be difficult, and I cannot say whether the present generalisation succeeds in providing the material for its solution."[25]

Embarking on the road to unification, the question for Einstein became how to choose between Weyl, Kaluza, and Eddington's theories. Although he found none of the theories satisfactory, he would borrow from them to construct his own models. All the while he would keep his eyes on the prize of describing the electron through a modified version of general relativity.

As 1919 drew to a close and the Roaring Twenties began, Einstein's life was fundamentally changed. He was now in his forties, well past the typical age of major achievements in theoretical physics. Yet his burning ardor to cap off his general theory of relativity with a unified description of forces and matter had only been ignited. He finally obtained a divorce from Mileva (ironically on Valentine's Day) on the condition that if he ever won the Nobel Prize, he would cede the money to her. Nothing could compensate her for her lost hopes, but the Nobel money would at least give her a chance for basic comforts.

After his divorce was final, Albert married Elsa on June 2. Months later, following Eddington's eclipse announcement, she realized that she had exchanged vows with the most famous scientist in the world. Elsa relished standing by her husband's side as he traveled around the world, met with celebrities, and received honor after honor.

Mileva's savvy request would pay off handsomely. Einstein would receive the 1921 Nobel Prize in Physics, accepting the award the following year. With her ex-husband evermore in the limelight, she and their two sons retreated from the public eye and lived off the prize money. Secure in his academic position, awash with fame, and ceding all domestic concerns to Elsa, Einstein was as free as an eagle to soar toward the lofty peaks of unification.

CHAPTER THREE

Matter Waves and Quantum Jumps

If you have to have these damn quantum jumps then I wish
I'd never started working on atomic theory.

—ERWIN SCHRÖDINGER (reported by Werner Heisenberg)

Please do not misunderstand me. I am a scientist, not a
teacher of morals.

—ERWIN SCHRÖDINGER, *Mind and Matter*

If lacking free will is like being in prison, general relativity is the ultimate jailer. By fusing time with space, it melds the past, present, and
future together into a solid block. Time's landscape is as frozen as a
Siberian gulag. All of history is forever locked in place; we just haven't
yet served our time.

Extending general relativity to include other forces would cement
our fate even further. A unified theory that explained electricity along
with gravity could in principle map out the neural connections of everyone who has lived or ever will live. We'd be condemned to think
thoughts and take actions that have forever been ordained. Once the
equations of eternity were set, our destinies would be sealed. As famously versed in *The Rubaiyat of Omar Khayyam*: "The Moving Finger writes; and, having writ, moves on: nor all thy piety nor wit shall
lure it back to cancel half a line. Nor all thy tears wash out a word
of it."[1]

Fate can be cruel. After the First World War ended, many of the
returning soldiers needed to heal their souls. While Schrödinger was

fortunate enough to have made it home safely, his beloved professor Fritz Hasenöhrl had been blown up by a grenade. Schrödinger and the Viennese academic community were stunned. In late 1919, Schrödinger's father died. Soon thereafter, the Austrian economy was decimated by stark inflation that wiped out many families' savings, including theirs. Times could not be harder. Schrödinger turned inward and began to think about his own life's direction.

Schrödinger found much emotional solace in female companionship. He kept a diary of the women with whom he had relationships. In that logbook he recorded meeting Annemarie "Anny" Bertel, a jovial, unpretentious woman from Salzburg, sometime in 1919. Although she was not an intellectual, she respected his bookish interests.

Unlike couples that fit together like a hand in a glove, Erwin and Anny were in some ways a mismatch. For example, they feuded about music—she loved playing piano, but he wouldn't tolerate it. While ultimately neither would see their relationship as exclusive, they always enjoyed each other's companionship. Thus it was a relationship based on familiarity and comfort. They soon became engaged and planned two weddings—one Catholic and the other Protestant—to respect the different faiths in their families. Both ceremonies took place in the spring of 1920.

During his postwar malaise, Schrödinger plunged into philosophy and became obsessed with the writings of Schopenhauer. In detailed notebooks, Schrödinger offered his comments and impressions of everything he read, describing Schopenhauer as the "greatest savant of the West."[2]

Stimulated by Schopenhauer's many references to Eastern philosophy, Schrödinger also delved into the Vedic writings of Hinduism (referring to them using the Sanskrit term "Vedanta") and other classics of Eastern thought. He briefly thought about switching his career to philosophy but decided to remain in physics and pursue the subject as a sideline. Over the years, he wrote several books expressing his own philosophical views, including *My View of the World*, based in part on a treatise he completed in 1925 called "Quest for the Path."

Schrödinger was particularly intrigued by Schopenhauer's explanation of passion and desire in the face of a mechanistic universe. Looking around him in the aftermath of the Great War, Schrödinger could see nothing but contrasts. While science and technology had leapt to unprecedented heights, culture, as he saw it, had sunk into

Dante-esque depths—what he called a "decay of the arts." "Our condition," Schrödinger noted, "bears a frightening resemblance to the final stage of the ancient world."[3]

Of course, given that he and Anny would maintain an open relationship throughout their married life, Schrödinger was hardly a puritan. But when he looked in the mirror he saw the modern-day equivalent of Plato or Aristotle—a polymath and Renaissance man—who happened to be trapped in a prurient age of decadence and violence.

In *The World as Will and Representation* and other works, Schopenhauer offered an explanation for the driving force of emotions that can lead to calamity. Drawing from the Hindu notion of karma and the Buddhist concept of suffering, he described how "will" is a universal force that compels people to carry out tasks. It is the desire that generates the action that brings about the inevitable. Just like other natural forces, it leads to predictable outcomes. However, the agents experiencing such compulsion fully believe it is their own volition producing the results. Neurotically, they can become immersed in their own longings, constantly feeling unfilled, because whenever a goal is reached a new desire bubbles up. Therefore, as Buddha noted, desire is suffering. One antidote is to eschew all goals and emotions and live an ascetic existence, something like a monk. Another possibility is to subsume your desire into aesthetic pursuits such as art or music. Instead of fruitless longing, write a stirring composition. But if you give in to desire, you should not be either condemned or praised, because you are just responding to a universal force.

Thus, if you fall in love with someone, it is not that you *chose* that person; rather, it is that your love is the agent carrying out the procedure of bringing you and the other person together according to your mutual destinies. From that vantage point, saying that Erwin and Anny chose each other has as much meaning as saying that Earth decided to pull the moon around it last month because of their fervent attraction to each other. He thereby saw no moral reason to abide by the traditional rules of marriage or to justify his impulsive decisions in general.

From that point on, Schrödinger wove philosophical motifs into his conceptual discussions of physics. The sense of wholeness that he gleaned from Schopenhauer's writings and the Vedic philosophy underlying those works would lead him to discount jumpy, incomplete descriptions of nature in favor of those with continuity, and fuzziness in favor of certainty. Ultimately, Schrödinger believed, everything in nature must be connected, flowing from one moment to the next in a continuous stream. (Note that he did explore the possibility of acausality in some of his writings, but the main thrust of his work favored causal connections.) Such considerations would end up playing a major role in his attitude toward the ambiguities of quantum mechanics.

The Heretic's Bible

There was considerable overlap—albeit with different emphasis—between Schrödinger's philosophical interests and Einstein's. Although Einstein too read Schopenhauer, he was much more strongly influenced by an earlier philosopher, Baruch Spinoza. Spinoza would be his guide to looking for a seamless, unified explanation of the universe, one in which chance played no fundamental role. Because Spinoza was one of Schopenhauer's key influences, Schrödinger read a great deal of Spinoza as well.

Spinoza was born in 1632 to a Sephardic Jewish family in Amsterdam. After an orthodox childhood in which he studied the scriptures, he developed a radical reinterpretation of God's role in the universe. The Sephardic community deemed his notion of the deity so heretical that it decided to excommunicate him—an exceedingly rare event in Judaism.

In traditional monotheistic religions, God plays an active role throughout history, starting with creating the world and bringing about life. As Creator, God is separate from the world, but He may choose to intervene whenever He wants. Not every decision is made by Him, however. He endowed humans with free will, so that they can make their own choices.

There are, of course, many theological differences regarding how often God intervenes and what the nature of human free will is. In faiths espousing predestination, humans' fates are sealed and their choices are preordained. Therefore, an evil person is doomed to make bad decisions—rendering "free choice" an exercise in showing why that individual is truly unworthy. In such a view, God's judgments and interventions have long been set in stone (perhaps forever), so that anything that happens was fated to occur.

In other faiths, choices are completely free, but a poor choice may lead to a punishing afterlife or perhaps bad luck later on in life. A good choice may make one feel closer to God and would likely be rewarded (though the manner depends on the particular faith). A personal God gazes down at what people do and reacts accordingly.

Starting in the seventeenth century, a more limited notion of God's intervention emerged in Europe, in which God's role is restricted to creating the universe, fashioning its laws, and stepping in only when needed to make adjustments. In that way, God behaves like a kind of clockmaker who creates His masterpieces and tinkers with them or resets them only as needed (the Great Flood being an example of a resetting of history). Newton ascribed to such a view, imagining God fashioning the law of gravity and other natural principles, setting the planets in place, and watching His beautiful creation proceed on its own—but reserving the right to intervene as needed to keep it running perfectly. The modern notion of a "miracle" involves the supposition that while occurrences stem from natural principles, sometimes God circumvents them to do good.

Spinoza's take on God and the universe was very unusual for his time. He rejected the notion of a personal God and the idea that God could selectively intervene in human affairs or in the natural world. Prayers, he believed, were a futile exercise because no one is listening. Rather, God is the substance that fills the universe itself—an infinite entity that pervades everything. All people and things are glimmering facets of a glorious, indestructible diamond.

Because, according to Spinoza, God is infinite and perfect, His nature is unchangeable. He has no choice whatsoever about how the universe takes shape, because its properties simply flow from His attributes. All events transpire from divine laws designed in the ideal way. Consequently, the history of the universe rolls out like a carpet that has been loomed with a timeless pattern. As Spinoza wrote in his *Ethics*: "In nature there is nothing contingent, but all things are determined from the necessity of the divine nature to exist and act in a certain manner."[4]

As Einstein moved away from the tangible and toward the ethereal—away from theories based on experimental questions and toward those shaped by abstract principles and aesthetic concerns—he began to evoke the name of God more and more in his statements about physics. This God wasn't the fatherly figure of the Bible, however, actively involved in human and worldly developments. Rather, it was Spinoza's deity—the perfect, timeless entity from which the laws of nature have sprung. As Einstein once responded to a rabbi's query as to whether he believed in God: "I believe in Spinoza's God who reveals Himself in the orderly harmony of what exists, not in a God who concerns Himself with fates and actions of human beings."[5]

In a much-discussed article that appeared in the *New York Times Magazine* on November 9, 1930, Einstein would name Democritus, St. Francis of Assisi, and Spinoza as the three greatest contributors in history to a "cosmic religious sense": a feeling of awe about the workings of the universe based on scientific investigation.[6] The naming of Democritus showed Einstein's belief in the importance of atomism. In the case of St. Francis, Einstein identified with his humanitarian concerns. Of the three, however, Spinoza was the maverick and the most controversial choice. Einstein's revelation of his views would spur much debate among religious scholars and clergy about the validity of "cosmic religion."

Einstein's belief in Spinoza's concept of cosmic order, perhaps along with his traditional Newtonian education in physics, led him to embrace strict determinism in his theories and reject any fundamental role for probability. After all, how could the rolling out of divine perfection happen in multiple ways? Every effect must have a clear cause, which in turn stems from an earlier cause, and so forth—a trail of fallen dominoes ultimately traceable to a supreme cause. His rejection of chance in quantum physics and his decades-long search for a seamless unified field theory would be clear ramifications of his zealous adherence to Spinoza's ideas.

A critical difference between the beliefs of Einstein and Schrödinger was the latter's devotion to Eastern thought. None of the figures Einstein mentioned in his piece about religion were from the Eastern tradition (he only briefly referred to Buddhism). He had little interest in any form of mysticism or spirituality. Schrödinger, on the other hand, had a deep sense that people share a common soul and that everything in nature is really a single entity. He distinguished this Vedantic belief in a kind of universal consciousness from Spinoza's view that humans are facets of the divine. The difference, Schrödinger emphasized, is that each of us is not a part, but rather the whole: "not . . . a piece of an eternal, infinite being, an aspect or modification of it, as in Spinoza's pantheism. For we should have the same baffling question: which part, which aspect are you? What, objectively, distinguishes it from the others? No, but, inconceivable as it seems to ordinary reason, you—and all other conscious beings as such—are all in all."[7]

Both Einstein and Schrödinger were driven to a quest for unity in science, but they had different motivations. For Einstein it was the search for the divine principles underlying nature—that is, the simplest, most elegant set of equations. For Schrödinger, it was looking for the commonalities in all things—a lifeblood running through the veins of everything in the cosmos. Because Einstein's belief was more rigid, he would never accept random elements as fundamental. Schrödinger would remain far more open-minded about randomness, seeing luck and chance as possible manifestations of universal will. Ironically, because of the power of will, a seemingly chance event could lead someone down a path he or she was meant to follow. Moreover, as he learned from his study of Boltzmann, the laws of thermodynamics derived from statistical averages of the sporadic behavior of myriad atoms. Billions of scattered droplets can produce a sea change.

Along with a striving for unity, a critical common element in Einstein and Schrödinger's scientific philosophies was a belief in continuity. Such a conception was grounded in the classical physics they grew up with, such as fluid mechanics, and enhanced by their shared sense—common to Spinoza's philosophy as well as Vedic philosophy—that events flow like a river from one moment to the next. Something couldn't simply vanish and reappear somewhere else or exert an instant unseen influence over a distance. Nature's garment must be sewn together by tight threads, both in time and space, lest it shred into a pile of tatters like a moth-eaten cloak.

Discontinuity was a hallmark of Bohr's planetary atomic model, which both Einstein and Schrödinger saw as a major weakness in a theory that was otherwise a key step forward. Why should electrons instantly jump from orbit to orbit in an atom when that never happens with planets in the solar system? "I can't imagine that an electron hops about like a flea," Schrödinger was known to say.[8]

Moreover, if electrons leap in atoms, why do they behave as a continuous stream in free space—within the empty interiors of cathode ray tubes, for example? Stimulated by the unification proposals of Weyl, Kaluza, and later Eddington, by the early 1920s Einstein had started to contemplate ways of explaining electron behavior through an extension of general relativity that included electromagnetism as well as gravity. The jumps, Einstein thought, must be mathematical artifacts of an otherwise deterministic continuous theory. Stimulated by discussions with Einstein, Schrödinger would independently develop his own idea of electron continuity, eventually resulting in his groundbreaking theory of wave mechanics.

Not everyone in the physics community saw discontinuity as a vice, however. While the rudiments of wave mechanics were taking shape, a pioneering young physicist from Munich, Werner Heisenberg, proposed an abstract mathematical theory called matrix mechanics in which instant jumps from state to state were de rigueur. Where else could such an abstract theory be proposed than in Göttingen's rarified environment? Heisenberg was inspired by a remarkable set of talks in that city by Bohr.

The Pathfinder's Quest

In June 1922, Hilbert and several other faculty members of the University of Göttingen, including Max Born, a bright young physicist, invited Bohr to deliver a series of lectures about atomic theory. By enthusiastically accepting the offer, Bohr broke an informal boycott against German academic institutions that had been in place since the First World War. Aside from Einstein, whose image was trumpeted internationally, the Germans' scientific reputation had greatly suffered because of the conflict. The ghastly repercussions of the German development of poison gas (by chemist Fritz Haber, a colleague of Einstein) and aerial warfare left deep psychological wounds among the survivors. Bohr's talks—dubbed the "Bohr Festival," after a recent "Handel Festival" held in the same city—helped open the door to renewed cooperation in science between Germany and other European nations.

It had been almost nine years since Bohr had first proposed his theory. In the intervening years, his proposal had been greatly enhanced by the contributions of Arnold Sommerfeld, working in Munich. In particular, Sommerfeld supplemented Bohr's enumeration of energy levels with two additional quantum numbers: total angular momentum and the component of angular momentum along one of the coordinate axes (usually taken to be the z axis). These allowed electrons with the same energy to orbit in different shapes and directions. The situation in which two states with different quantum numbers have the same energy is called degeneracy.

Degeneracy is something like tossing a bunch of horseshoes at a stake and having all of them land so that they are leaning directly on it, but at different angles. Because all of them would be touching the stake, they'd be counted as equal, despite the variation in how each horseshoe is tilted. Similarly, electrons in degenerate states have equal energies but different tilts and shapes to their orbits.

In 1916, Sommerfeld, along with Dutch chemical physicist Peter Debye, demonstrated that his enhanced version of the Bohr model, known as the Bohr-Sommerfeld model, could explain an enigma called the Zeeman effect. First observed by Dutch physicist Pieter Zeeman in 1897, the effect involves placing a gas of identical atoms in a magnetic field and observing the spectral lines produced. As the magnet is turned on, some of the spectral lines split up. Instead of one line at a certain frequency, suddenly there are three, five, or even more around that frequency. It is like tuning into a certain radio station that dominates its frequency range, scanning for new stations, and unexpectedly finding two more with close (but not exactly the same) frequencies.

Sommerfeld showed how the Zeeman effect was the result of interactions between the applied magnetic field and the angular momentum of the electrons orbiting the atomic nucleus. These tugs of the magnetic field make it so that orbits with different angular momenta, instead of being degenerate and having the same energy, instead have slightly different energies. Because different energy levels lead to different frequencies of the light given off as electrons hop from state to state, the split in energies causes the divergence in spectral lines.

Sommerfeld was fortunate to have two brilliant physics students who would each go on to make a mark in quantum theory. One of them was Wolfgang Pauli, the Viennese godson of Mach. He was a true wunderkind, impressing older physicists with his precocious insights. At the tender age of twenty, when he had been a university student for only two years, Sommerfeld invited Pauli to contribute a review article about relativity to an encyclopedia of mathematical sciences that he edited. Pauli complied, writing a masterly summary of the topic. Pauli became known not only for his erudition and quick take on subjects but also for his brutal directness. He felt obliged to tell his colleagues his honest opinion of them and their research even if his comments sometimes dug in like a knife. He called, for example, Sommerfeld's numerical theories about atoms "atomysticism."

The other quantum virtuoso trained by Sommerfeld in the early 1920s was Heisenberg. Heisenberg was a strapping young man, equally at home with pencil and paper or tramping along rugged mountain paths. He came into Sommerfeld's group as a member of the Pathfinders (in German, *Pfadfinder*), a German equivalent of the Boy Scouts that had, at the time, strong nationalist elements.

Heisenberg had a deep respect for Einstein and was fascinated by relativity. He was impressed and delighted whenever Sommerfeld read one of Einstein's letters out loud during class. However, Pauli helped persuade Heisenberg not to pursue studies in that area. After finishing his encyclopedia article, Pauli was convinced that there weren't many basic problems yet to be solved in relativity that could be easily tested by experiment. Therefore, relativity, as Pauli saw it at the time, was not ripe for progress. The real hot area, he advised Heisenberg, was atomic physics and quantum theory.

"In atomic physics we still have a wealth of uninterpreted experimental results," Pauli explained to Heisenberg. "Nature's evidence in one place seems to contradict that in another, and so far it has not been possible to draw an even halfway coherent picture of the relationship involved. True, Niels Bohr has succeeded in associating the strange stability of atoms with Planck's quantum hypothesis . . . but I can't for the life of me see how he could have done so, seeing that he too is unable to get rid of the contradictions I have mentioned. In other words, everyone is still groping around in a thick mist, and it will probably be a few years before it lifts."[9]

In the summer of 1922, Einstein was invited to give a talk in Leipzig about general relativity. Sommerfeld strongly encouraged Heisenberg to attend and offered to introduce him to Einstein. Heisenberg was thrilled. However, anti-Semitic threats against Einstein caused him to cancel and send Max von Laue in his place. Not knowing that Einstein wasn't there, Heisenberg traveled to Leipzig's assembly hall anyway. He was startled to see students of a Nobel laureate in physics, Philipp Lenard, standing in front of the hall handing out red leaflets denouncing Einstein and relativity, claiming it to be "Jewish science." Lenard had started an anti-Semitic campaign to eliminate any form of science that wasn't "purely German." Little did Heisenberg know at the time that Lenard's credo would, in less than a decade and a half, become national policy under the Nazi regime.

Another speaker Sommerfeld urged Heisenberg to see was Bohr. They decided to go together to Bohr's event. Attending the Bohr Festival was kind of a homecoming for Sommerfeld, who had received his doctorate at Göttingen. By then, Pauli was also at that university, serving as Born's research assistant in a postdoctoral position. After a pleasant journey to Göttingen, Sommerfeld and Heisenberg took seats in the crowded lecture hall to hear Bohr speak.

It was a glorious season for Göttingen to roll out its welcome mat for the international scientific community. Beautiful, sunny summer weather brought out the charm of the city, with its medieval buildings, market stands, and streetcars. Gorgeous flowers lined the paths leading up to the auditorium. Joviality and excitement filled the hall as the Bohr Festival began.

Bohr's lecture style was not for casual ears. He was very soft-spoken and often used abstruse, enigmatic language. However, in some ways these difficulties added to his mystique as a kind of high priest of quantum theory. Just like the Delphic oracle, who spoke cryptically, Bohr's inscrutable style of lecturing allowed audience members to construct their own interpretations. For example, although Bohr never explicitly explained the physical principles behind his angular momentum quantization rule, many physicists assumed that it must have had a logical origin and that he had some way of justifying it through classical mechanics.

Heisenberg, though, wasn't so easily satisfied. As he listened intently to the lecture, he started to doubt that Bohr had thought his theory through completely. When it came time for the question period, he shocked many of the professors in the audience by confronting Bohr about the differences between the classical and quantum idea of orbital frequencies. In Bohr's model, Heisenberg pointed out, the frequencies of electrons have nothing to do with their rates of orbit. Could Bohr justify that? Also, Heisenberg wondered if Bohr had made any progress studying atoms with multiple electrons. Was his theory still only applicable to hydrogen atoms and single-electron ions?

The audience, no doubt, was stunned by Heisenberg's comments. At that time, it was almost unheard of for a student to raise questions in a public talk about a professor's theory, let alone challenge the internationally famous Bohr. Bohr took the comments in stride and invited Heisenberg to go for a long walk with him in the nearby hills to discuss the issue. Bohr confided during the stroll that aspects of his theories were based on intuitive ideas rather than physical principles. Heisenberg was gratified that such a noted thinker would reach out to him so warmly. It would be the first of countless walks that they would take together to mull over the philosophy of the quantum.

Reality's Matrix

Heisenberg's interactions with Bohr helped inspire him to develop his own theory of atomic transitions. After all, if Bohr didn't have all the answers, the field was ripe for a more comprehensive vision of the atom. Working without preconceptions, Heisenberg was unafraid to set aside widely accepted beliefs, such as the notion that quantum numbers must be whole numbers.

Using spectral data that he had obtained from Sommerfeld, Heisenberg had earlier constructed a system called the "core model" that used half-integer as well as integer quantum numbers. The half-integers helped explain doublets: spectral lines that appeared in pairs. Sommerfeld had snappily dismissed Heisenberg's hypothesis, advising him that quantum numbers $\frac{1}{2}$, $\frac{3}{2}$, and so forth were "absolutely impossible." Bohr had similarly rejected the idea. However, Heisenberg's ideas would resonate with Born, with whom he would have a chance to collaborate.

As a youthful faculty member of a university known for challenging convention, Born was open to radical suggestions. He had been tinkering with alternatives to the Bohr-Sommerfeld model himself. As fate would have it, during the 1922–1923 academic year Sommerfeld took a leave to travel to the United States and teach at the University of Wisconsin. During his absence, he dispatched Heisenberg to Göttingen to work with Born. Completing what had become a quantum triangle of Munich, Göttingen, and Copenhagen, Pauli shifted north to become Bohr's assistant.

When Heisenberg arrived in October 1922, Born advised him to focus on variations of Bohr's theory based on principles from astronomy and orbital mechanics. They worked together on trying to match planetary models with the spectral lines of ionized helium (helium with a single electron), the simplest system beyond hydrogen.

In May 1923, Heisenberg headed back to Munich to complete his doctoral program and his final oral defense. Despite Sommerfeld's excellent theoretical contributions, the emphasis there was still on the practical side of physics. Unlike Schrödinger, Heisenberg had little experience with or inclination toward experimentation and ended up doing very poorly in that part of the defense. His marks for the theoretical and experimental sections averaged to something like a C. Nonetheless,

Sommerfeld still held a PhD party in his honor. Embarrassed by the mediocre score, Heisenberg left the party early, bolted to the railway station, and hopped on the midnight train to Göttingen to resume his collaboration with Born, this time as a paid research assistant.

There was much for Heisenberg to do. New data about spectral lines were pouring in, with curious patterns suggesting increasingly intricate structures, necessitating more and more changes to existing models. Heisenberg tried in vain to adapt his core model to the novel data.

By early 1924, Born began to realize that their efforts to apply a planetary analogy to electrons had failed. Traditional orbital mechanics, combined with quantized energies and angular momenta, simply couldn't explain how electrons in ionized helium behaved. If helium, a relatively simple system, couldn't be modeled, what hope was there for understanding all the complex atoms that made up the periodic table?

Tossing classical mechanics aside when it came to atoms, Born declared the need for a wholly novel "quantum mechanics." The key difference was that quantum mechanics would be discrete rather than continuous, operating on the basis of instantaneous jumps, rather than smooth transitions. Mapping out the behavior of electrons would thereby require treating the atom like a black box with cloaked inner workings, rather than as a classic physical system.

Born's move was unprecedented in the history of physics. Since the time of Newton, physicists had treated the laws of motion as sacrosanct. Einstein's special theory of relativity had modified the definitions of momentum and energy but not changed the basic premise that these quantities are strictly conserved (by including relativistic mass as another form of energy) and that nothing simply vanishes somewhere and reappears elsewhere. In Newtonian physics, every instant of time must be accounted for; hidden moments happen experimentally but not theoretically. Born could well have said that we don't understand the mechanism for electron jumps because of observational limits or the noise created by the interference of complex processes. Instead, he surgically excised any causal connections between an electron's situation before and after a leap. All that could be known are transition rules.

If classical mechanics was something like a miser who tracks every penny of his savings at every moment, quantum mechanics presented like a mutual fund customer who just cares about the prospects for his money to grow. If he even bothered to inquire about his investment, he'd be told, "Don't ask; it just happens." Likewise, in quantum mechanics there is no direct mechanism for electron jumps; they just follow a rule book involving initial and final states.

Similarly frustrated by the limitations of classical mechanics, Heisenberg was primed for a wholly new approach. Throughout 1924 and early 1925—part of which time was spent as a visitor to Bohr's Institute in Copenhagen—he tinkered with various ways of matching orbital electron behavior to complex spectra. After consulting with Pauli, Bohr and others, Heisenberg decided to abandon the idea of describing electron orbits. Instead of trying to visualize the paths that electrons take, he felt it would be more productive to focus solely on quantities that could be measured directly, known as observables.

A breakthrough came in June 1925, when Heisenberg enjoyed two weeks of uninterrupted thought on the island of Heligoland in the North Sea. Severe hay fever had brought him to that refuge; the sea air offered him relief from the sniffles. There he developed a system for calculating the amplitudes (related to the likelihood) of jumps between electron states that would produce particular frequencies of emitted or absorbed light. He developed a kind of spreadsheet that tallied these amplitudes for all possible atomic transitions. He also showed how mathematical operations based on these tables could be used to determine the probabilities that electrons would have certain positions, momenta, energies, and other observable quantities. Hence, such physical quantities would be known not exactly but probabilistically, like the odds of getting twenty-one when dealt a hand of cards in the game of blackjack.

Returning to Göttingen, Heisenberg presented his table of amplitudes to Born, who soon recognized it to be a type of matrix: a mathematical entity with numbers arranged into rows and columns. Born recruited one of his doctoral students, Pascual Jordan, to work with Heisenberg and him in exploring the mathematical implications of what became known as "matrix mechanics."

Born was well familiar with the property of matrices that multiplying two of them together produces different answers depending upon their order. Unlike standard multiplication, in which 2 × 3 is the same as 3 × 2, in matrix multiplication A × B is not generally the same as B × A. If the order does not matter, the quantities are said to "commute"; those that depend on order are called "noncommutative." Because in Heisenberg's system noncommutative matrices are used to determine physical properties such as position and momentum, the order of operation matters for such measurements. Thus if the position of a state is measured first and then its momentum is measured, the result is different than if the momentum is assessed first and the position second.

Heisenberg would later show that this noncommutivity leads to an "uncertainty principle" that makes exact simultaneous measurement of certain paired quantities impossible. For example, the position and momentum of an electron cannot both be known precisely at once. If one is pinned down, the other must be murky. It is like a photograph in which either the foreground or background can be in perfect focus, but not both. If the photographer tries to sharpen an image in the foreground, the background would become shadowy; the converse is also true. Similarly, if a physicist constructs an experiment designed to reveal the pinpoint location of an electron, its momentum would become smeared out over an infinite range of values; that is, it couldn't be known at all.

The abstraction of matrix mechanics did not endear it to the experimental physics community, with its bent toward the tangible. Only after its sister theory of wave mechanics appeared and the two were shown to be equivalent did the united theory of quantum mechanics become widely accepted.

Einstein's belief in Spinoza's clockwork deity drove him to recoil from one of the startling implications of Heisenberg's theory: if position and momentum can never be measured simultaneously and precisely, then it is impossible to plot out the locations and velocities of all things in the universe and predict their future development. Such an omission did not bother Heisenberg and Born, who had become comfortable with a probabilistic mechanics instead of an exact classical mechanics. Einstein battled vehemently against dropping such strict determinism in favor of a kind of particle roulette.

Counting Photons

It is curious that Einstein, one of the fathers of quantum theory, would flee from his own creation. However, we must distinguish the original idea of a quantum, which simply means a discrete unit of energy or other physical entity, from full-blown quantum mechanics, a system that replaces deterministic classical mechanics on the atomic scale. In Einstein's photoelectric effect, for example, an electron absorbs a discrete amount of energy in the form of a photon, but then uses the boost to escape the surface of a metal and move continuously (and deterministically) through space. Einstein objected to the contrasting idea that an electron would gulp down a photon and then instantly vault to a completely different location. Seemingly discrete, random jumps must have continuous, causal explanation through a deeper theory, Einstein surmised.

Einstein saw no problem with randomness as a tool rather than as a fundamental aspect of nature. Statistical mechanics, Einstein knew, needed randomness as a way to account for the aggregate behavior of myriad atoms interacting with each other and with their environment. Classical mechanics adeptly handles simple interactions between pairs of objects but falls short in addressing complex systems with large numbers of components. That's where chance comes in, Einstein believed—not as a basic element but rather as a way of representing a hodgepodge of motions.

Einstein's last major contribution to quantum theory before he switched camps and became its best-known critic was a quantum statistical theory of ideal gases. An ideal gas is a large set of molecules, generally housed in a container, that for simplicity's sake are presumed not to interact with each other. In classical statistical mechanics, developed by Boltzmann and others, assumptions of random motion leads to a simple relationship between pressure, volume, and temperature, called the "ideal gas law." Einstein updated standard statistical mechanics to encompass the notion that energy is quantized.

The impetus for Einstein's final venture into the quantum realm was a remarkable paper he received from Indian physicist Satyendra Bose that derived Planck's blackbody radiation law from quantum statistical principles. Einstein translated the paper into German, and published it in the August 1924 issue of the prestigious journal *Zeitschrift für Physik*. Bose looked at photons as something like identical ping-pong balls in a container, carrying discrete energies that (following Planck's law) depend on their frequencies. Einstein generalized Bose's idea to monatomic gases (those containing a single type of atom). Therefore, the quantum statistics of certain types of identical particles, including photons, is known as Bose-Einstein statistics. (That's where the term "boson," recently applied to the Higgs particle, arises.)

In September 1924, one of the most important scientific conferences of the inter-war years, the *Naturforscherversammlung*, took place in the beautiful Alpine city of Innsbruck, Austria. While Einstein did not present a talk at the meeting, he attended its sessions and had the opportunity to discuss his ideas on quantum statistics informally with various participants, including Planck.

Schrödinger also attended the conference. It offered him the opportunity to meet Einstein and Planck, two of the physicists he most respected—and, of course, two of the most famous physicists in the world. He had seen Einstein lecture at the 1913 Vienna meeting and had exchanged articles with him about general relativity, but up until that point he had not spoken with him—at least not in depth.

The meeting of Einstein and Schrödinger in Innsbruck would prove not only the start of their long, fruitful friendship (beginning formally but warming over time) but also a key juncture in the history of modern physics. Einstein's valedictory efforts in the area of quantum statistics, discussed at the meeting, would inspire Schrödinger to engage him in correspondence and eventually learn from him about French physicist Louis de Broglie's notion of matter waves. That lead, in turn, would stimulate Schrödinger to construct his own wave equation, one of the key pillars of quantum mechanics.

In Innsbruck, Schrödinger also welcomed the chance to catch up with his Austrian colleagues—as he was then working in Switzerland—and breathe the clean mountain air. The latter was important because three years earlier he had developed lung problems due to a severe bout of bronchitis followed by a case of tuberculosis. He also smoked heavily, which didn't help his breathing.

The past few years had been turbulent for Schrödinger in general. After he married Anny, he had become quite a vagabond scholar. Although offered a position at the University of Vienna, he decided to move in succession to the German cities Jena, Stuttgart, and Breslau (the last now the Polish city Wroclaw) for brief academic positions at each from late 1920 until late 1921. Salary was a great concern for him, as inflation was beginning to ravage Germany. He watched with horror as his widowed mother, once proudly middle-class, lost her home and lived in squalor after his father's death. She passed away from cancer in September 1921. Erwin was determined to land the most lucrative and secure academic position he could find, hoping to offer Anny a comfortable lifestyle and spare her from any chance of poverty.

Such an opportunity arose late that year when a position opened up at the University of Zurich. Switzerland offered Erwin and Anny a stable, peaceful environment, free of the economic problems and unrest facing Germany and Austria. Once settled in the position, and after dealing with the aforementioned bronchitis and tuberculosis attacks, he began to publish articles extending the classical ideas of Boltzmann to the quantum domain.

One question that concerned Schrödinger during his early years in Zurich was defining in quantum terms the entropy, or amount of disorder, for an ideal gas. Boltzmann defined entropy in terms of the number of unique microstates (particle arrangements) for each macrostate. However, if particles are indistinguishable, such as in a quantum gas, there are fewer unique states. It is like counting the number of arrangements of a set of pennies, each minted in a different year. If you distinguish them by their dates, they have many more unique configurations than if you treat them as identical. Therefore, quantum estimates of entropy are different from classical measures.

Before Bose contributed his seminal paper on photons and Einstein extended his treatment to include ideal gases, many physicists were perplexed about which factors to include in expressing the entropy for quantum systems. A well-known equation for entropy contained a controversial correction term that no one, until Bose, could fully explain. The correction term was added to rectify problems with Boltzmann's formula as applied to quantum gases. But not everyone believed in its validity. Schrödinger published a 1924 paper that left out the correction term and led to what turned out to be an erroneous expression for entropy.

Given Einstein's newfound methods, Schrödinger's encounter with him in Innsbruck and their subsequent correspondence were eye-openers. Einstein's insights inspired Schrödinger to think about quantum statistics in a whole new way, setting aside his classical misconception that rearranging particles always leads to different microstates. It took a little while, though, for the implications to sink in. At first Schrödinger thought there must have been a mistake in Einstein's calculations, because they disagreed with Boltzmann's methods. His initial letter to Einstein, in February 1925, pointed out the supposed error. Einstein patiently responded by explaining Bose's idea that photons can share identical quantum states. Schrödinger revised his definition of entropy based on the new statistics and presented his work to the Prussian Academy of Sciences in July 1925.

A theorist cannot anticipate which section of a research article might prove the most stimulating. Sometimes even a tangential mention can trigger the imagination and unleash a cascade of fruitful ideas. A reference in one of Einstein's quantum statistics papers to the work of de Broglie would inspire Schrödinger to develop his greatest contribution to science—the Schrödinger equation of wave mechanics. As physicist Peter Freund pointed out, "Without Einstein's endorsement of de Broglie's work, the Schrödinger equation might have been found quite a bit later."[10]

Matter Waves

A particle and a wave seem to be entirely different things. One is concentrated and the other is spread out. One bounces off walls and the other slinks around corners. One seems to be a tiny part of matter and the other presents itself as ripples through space. What could they possibly have in common?

Photons, as first shown by Einstein, represent a hybrid of both. Like particles, they carry parcels of energy and momentum, which they can dole out in collisions. Like waves, they have peaks and valleys, which can line up with each other in the striped images called interference patterns.

In his 1924 doctoral thesis, based on calculations completed the previous year, de Broglie imaginatively applied such duality to everything. Not just photons, he hypothesized, but all types of substances have both particlelike and wavelike aspects. In particular, an electron wriggles with a wavelength that is Planck's constant divided by its momentum.

The beauty of de Broglie's concept is that it naturally leads to Bohr's quantization of angular momentum (and to a generalization by Sommerfeld called the Bohr-Sommerfeld quantization rules)—the key to stable orbits. De Broglie imagined an electron's orbit in an atom to be something like a plucked guitar string, only circular. Just as a guitar string can vibrate in different modes, with various numbers of peaks and valleys, an electron wave in an atom similarly can undulate with various wavelengths. Because momentum, in de Broglie's formula, is inversely proportional to wavelength, and angular momentum is momentum times radius, that leads to a rule that confines angular momentum to discrete values. Thus a simple calculation produces the critical constraint on electrons that Bohr couldn't adequately explain himself, yet which is all-important to his theory.

In one of his papers on the quantum statistics of monatomic gases, Einstein drew upon de Broglie's matter wave idea as an explanation for how, in a low-temperature gas, atoms move in lockstep, thus making them more orderly and decreasing their quantity of entropy. The notion that atoms, like photons, could behave as waves drew a vital connection between the atomic gas of Einstein and the photon gas of Bose, upon which Einstein based his theory. Einstein also praised de Broglie for his innovative solution to the angular momentum quantization problem, which had been an embarrassing gap in Bohr's model.

When Schrödinger perused Einstein's paper and found the reference to de Broglie's thesis, he itched to get his hands on it as soon as possible. Ironically, he didn't seem to realize that its major results had already been published and had been available for some time at the University of Zurich library, right under his nose. Instead he wrote to Paris and obtained the actual dissertation. Primed by his reading of Schopenhauer and Spinoza to look for unifying principles, Schrödinger found his imagination stirred by the brilliance of de Broglie's thinking on common aspects of matter and light. Suddenly the Bohr-Sommerfeld model of the atom transformed from a flawed solar system analogy to a pulsating, beating heart of matter, throbbing in natural patterns that determined its properties. On November 3, 1925, Schrödinger wrote to Einstein, "A few days ago, I read with the greatest interest the ingenious thesis of Louis de Broglie, which I finally got hold of."[11]

Encouraged by Debye, who was then at ETH, Schrödinger offered a colloquium about de Broglie's matter waves. The seminar wonderfully showed the revolutionary implications of the idea. At the end of the talk, Debye suggested that Schrödinger investigate what kind of equation could model such waves, showing how they developed over time and space. Just as electromagnetic waves are explained by Maxwell's equations, could there be a mechanism to produce matter waves that match the physical constraints of any given situation? For example, how would electrons behave when they are subject to the electromagnetic field created by protons in atomic nuclei? How would they behave outside of atoms as they moved through empty space?

Schrödinger spent the next few months in a frenzy trying to find the right equation that would generate matter waves and explain the behavior of electrons, both in and out of atoms. Frustrating his early efforts was an intrinsic property of electrons that hadn't yet been recognized, called "spin." First identified in 1926 by two of Ehrenfest's students, Samuel Goudsmit and George Uhlenbeck, spin is a quantum number that expresses the behavior of a particle in an external magnetic field. Spin "up" means that the particle is aligned in the same direction as the field, and spin "down" means that it is situated in the opposite direction. Many types of particles, including electrons, possess spin values that are half-integer quantities, such as ½ or –½. These half-integer spin particles do not obey Bose-Einstein statistics, because they cannot share the same quantum state. Rather, as Pauli proposed, electrons and other half-integer spin particles must obey an "exclusion principle" mandating that each must occupy its own quantum state. Particles of that type, now called fermions, cannot bunch up, like concertgoers in a mosh pit. Rather, each has its own seat.

The term "fermions" derives from Fermi-Dirac statistics: the proper description for the collective behavior of particles of half-integer spin. Named after Italian physicist Enrico Fermi and British physicist Paul Dirac, who each contributed to the theory, it tallied particle states differently than Bose-Einstein statistics did. Dirac would later come up with the correct relativistic equation for fermions, called the Dirac equation. It would require a new type of notation involving complex numbers.

Schrödinger began his calculations without knowing all this, and soon developed an equation for matter waves that made use of special relativity. It was a solid, important equation, later rediscovered by Swedish physicist Oskar Klein and German physicist Walter Gordon and called the "Klein-Gordon equation." The trouble was, it didn't quite apply to electrons and other fermions, because of their half-integer spin. (It would work for spinless bosons, but he was trying to describe electrons, not bosons.) To his great disappointment, when he tried to model the Bohr-Sommerfeld atom, his predictions were off the mark.

After flogging the dead horse for some time, Schrödinger decided that he needed a break. The Christmas holidays were coming up, and it would be the perfect time to get away and think deeply about matter waves. He let Anny know that he would be heading to a villa in the scenic Alpine village of Arosa, Switzerland. The village was familiar to him because he had convalesced there after his lung infection. Meanwhile, he wrote to one of his ex-girlfriends from Vienna (whose name is unknown to history, as his diary from that year is missing) and invited her to join him. Anny would stay behind in Zurich.

Christmas Miracle

In his personal, philosophical essay "Quest for the Path," completed in 1925, Schrödinger identified with Schopenhauer's idea of will as a shared force that drives all people and things toward their destinies. He used the analogy of a sculpture to show that while the end product is solid, beautiful, and timeless, it takes thousands of tiny, apparently haphazard and destructive whittles to a stone in order to produce it. "At every step we have to change, overcome, destroy the form which we have had hitherto," Schrödinger wrote. "The resistance of our primitive desires, which we encounter at every step, seems to me to have its physical correlate in the resistance of the existing form to the shaping chisel."[12]

Schrödinger was proudly impulsive, feeling that risks were essential to growth. As the world bid adieu to 1925, he checked his old girlfriend and himself into the Villa Herwig, surrounded by gorgeous mountain scenery, preparing himself for an intense period of calculations. Whatever he did there seemed to work, as the two-week holiday would inaugurate the most productive period of his life, generating an entirely new approach to physics that would gain him a Nobel Prize. As Hermann Weyl, who knew the Schrödingers well and apparently had inside word about the liaison, described the period to scientific historian Abraham Pais: "Schrödinger did his great work during a late erotic outburst in his life."[13]

The "late" refers to the fact that Schrödinger was thirty-eight at the time of his prolific interval, far older than the boy geniuses Heisenberg and Pauli, who commanded the other quantum flank. Sad to say, few theorists (at least in modern times) make major contributions in their late thirties or older. Einstein offers another exception to the rule; he completed general relativity at the age of thirty-six and his contributions to quantum statistics at the age of forty-five. However, unlike Schrödinger, his Nobel citation was for work completed in his twenties (the photoelectric effect), not his thirties.

Primed with an unexpected burst of youthful energy, Schrödinger barreled toward his destination. After continuing to toy with a relativistic wave equation, he decided to switch to a nonrelativistic version. Instead of $E = mc^2$, he invoked the older Newtonian formula for energy. Combining the classical expression for kinetic energy (energy of motion) with that of potential energy (energy of position), he cleverly rewrote these as a mathematical function called the Hamiltonian operator (similar to Hamilton's formulation mentioned earlier, but expressed in terms of derivatives and other functions). In what is now a famous equation, Schrödinger applied the Hamiltonian to an entity called the wavefunction (also known as the psi function) and demonstrated how the former transformed the latter.

A wavefunction, according to Schrödinger's conception, represents how an elementary particle's charge and material are spread throughout space. To find the stationary states of a particle with fixed energy—for example, the stable electron states of an atom—simply look for all the wavefunctions for which application of the Hamiltonian produces a number times the wavefunction. Each number for which that equation is true represents an energy level, and each wavefunction represents the stationary state corresponding to that energy level.

Let's use a basic analogy to understand how Schrödinger's method works. Suppose you are a banker living in a country where there are many counterfeit banknotes. You develop a scanner that looks for the real notes by looking for a number in one of their corners that indicates their true value. If a note doesn't have that number, it is declared counterfeit and worthless. On the other hand, if the scanner does detect the number, an indicator lights up with the note's value and the note is placed in one of several piles according to that worth. Think of the Hamiltonian, then, as a scanner that processes wavefunctions and in some cases reads out their energy value and keeps them, while in other cases it discards them. The mathematical terms for the results of such a sorting process are called "eigenvalues" (the term means "proper values") and "eigenstates" (proper states). The Hamiltonian applied to an eigenstate (the stationary state wavefunction) yields an eigenvalue (the energy) times that eigenstate.

Schrödinger was, of course, keen to solve the hydrogen atom problem with his new method. He noted that the electric field of an atomic nucleus radiated outward in all directions, which gave the problem a kind of spherical symmetry. Exploiting that symmetry, he produced a set of solutions that could be classified by three different quantum numbers—precisely those proposed by Bohr and Sommerfeld. To his delight, his revised formula, now listed in every modern physics textbook as the Schrödinger equation, yielded the correct results, marvelously reproducing the Bohr-Sommerfeld atom.

By the end of January 1926, Schrödinger's first paper on the topic, "Quantization as an Eigenvalue Problem," was finished. Completing such an important breakthrough in only a couple of months was a virtually unprecedented feat. He sent a copy to Sommerfeld, who was staggered by his brilliant accomplishment. Sommerfeld replied that the paper hit him "like a thunderclap."[14]

Given his great respect for Planck and Einstein, Schrödinger waited eagerly for their reactions as well. Fortunately, these were largely positive. As Anny recalled: "Planck and Einstein were very, very enthusiastic from the start. . . . Planck [said], 'I'm reading it like a child reads a puzzle.'"[15]

Schrödinger thanked Einstein in a personal note. "Your and Planck's approval are more valuable to me than half the world. Besides, the whole thing . . . possibly never would have existed (not from me at least) if your work did not make obvious to me the importance of de Broglie's ideas."[16]

By then, several papers by Heisenberg, Born, and Jordan outlining the theory of matrix mechanics had already been published. Dirac had also developed a clever mathematical shorthand for describing quantum rules using bracket symbols that rendered matrix mechanics much more elegant and straightforward. The natural question arose as to the connection between wave mechanics and matrix mechanics, given that each deftly handled the hydrogen atom question but in a different way. Schrödinger was careful to emphasize that his theory had been independently developed and was not based at all on Heisenberg's work.

Despite the independent origins of his and Heisenberg's theories and his natural preference for the former, Schrödinger realized the importance of demonstrating their equivalence. Sommerfeld sensed right away that the theories were compatible—but that compatibility needed to be proven mathematically. Schrödinger soon supplied the proof, which was seconded by an even more rigorous proof by Pauli. With both theories shown to be equally valid, Schrödinger began to argue that his was more tangible and physically reasonable. After all, it described how electrons behave in space and time, rather than how they transform in an abstract world of matrices.

In the Realm of Ghosts

Born thought hard about the implications of both theories and started to see cracks in each, even the one he helped develop. He was well aware of the criticism that matrix mechanics was too abstract. The wave approach was indeed more concrete and visualizable. It modeled well processes that happened in real physical space, such as collisions. Born had to concede its elegance, clarity, and value.

On the other hand, wave mechanics offered an untenable vision of electrons being distributed over entire regions of space. Such a picture did not match experimental observations that showed them behaving sometimes as point particles. As vivid as the image of an electron oscillating in space seemed, there was no observational evidence that its material and energy were actually spread out.

To reconcile the two approaches, Born proposed a third way: to imagine the wavefunction as a "ghost field" that guides the true electron. The wavefunction would possess no physical characteristics of its own—neither energy nor momentum. It would reside in an abstract space (now called Hilbert space) rather than real space, making its presence known only indirectly when electrons were observed by supplying information about the likelihood of certain outcomes. In other words, it would serve in similar fashion to Heisenberg's state matrix as a reservoir of data about probabilities.

Born demonstrated how different observable quantities could be found using the wavefunction in its ghostly, "behind the scenes" role. Every time a measurement was taken, the probabilities for different outcomes would depend on the eigenstates of a particular operator (mathematical function) applied to that wavefunction. For example, to measure the most likely location of an electron, find the eigenstates of the position operator and use these to calculate the probabilities of every possible position. To find its most likely momentum, do the same with the momentum operator and the momentum eigenstates. A precise measurement of either position or momentum meant that the electron wavefunction matched one of the position or momentum eigenstates, respectively. The strange thing was, because the position and momentum eigenstates formed different sets, you could never measure position and momentum simultaneously. You needed to choose an order: either position first or momentum first. As with matrix mechanics, the order of operations produced different results.

You could also use wavefunctions, in Born's interpretation, to determine the likelihood that an electron would convert from one quantum state to another—for example, to leap suddenly between two energy levels of an atom. Such a "quantum jump" would be instantaneous and unpredictable, except for its odds of happening. The only way to see the jump would be to observe the impact on the atom's spectrum, either the release or absorption of a photon. You wouldn't actually observe the electron's movement through space.

In short, Born's approach transformed Schrödinger's wavefunctions from physical waves into probability waves. In their updated role they could tell you only how likely it would be that electrons would have certain positions or momenta and what were the odds that these values would change; you could never pin down both values at the same time. Because you would never know at any given moment both where a particle is and how it is moving, you could never predict exactly where it would be the next instant. Thus Born turned Schrödinger's deterministic description into a probabilistic, indeterminate series of quantum jumps from state to state.

Heisenberg strongly agreed with Born that electrons could not literally be waves spread throughout space. The only point he saw to wave mechanics was to provide an alternative means of calculating the matrix components of his own theory. To imagine electrons as kinds of undulating blobs surrounding atomic nuclei seemed ridiculous to him. No quantum experiment showed electrons as distended objects. Therefore, he welcomed Born's interpretation as a way of distilling the useful results from Schrödinger's calculations while discarding the hokum of bloated electrons.

The House of Bohr

Matters came to a head in October 1926 when Schrödinger visited Copenhagen, at Bohr's invitation, to present his results. Bohr's Institute for Theoretical Physics had become a holy temple of quantum pontification, with Bohr as the chief pontificator. Surrounding Bohr was an enthusiastic cadre that included (at the time) Heisenberg, Dirac, and Oskar Klein.

Klein was particularly interested in wave mechanics, as he had developed his own take on the subject. He too had read de Broglie and wanted to construct a wave equation based on the idea of matter waves. Trying several different approaches, he independently developed a form of Schrödinger's equation in late 1925 but, because of illness, never had the chance to submit it for publication. By the time he was well, Schrödinger's first paper had already appeared. Klein did get credit, however, along with Gordon, for the relativistic version of the equation.

Klein also independently reproduced Kaluza's theory of extending general relativity by an extra dimension with the goal of bringing electromagnetism and gravity under the same banner. Like his predecessor, Klein hoped to develop a unified theory of nature that would explain how electrons move through space under a combination of forces.

In contrast to Kaluza's theory, though, Klein's was grounded in quantum principles. It made use of de Broglie's notion of standing waves, but treated them somewhat differently. Instead of being wrapped inside an atom, the waves were curled around an unseen fifth dimension. Klein equated the momentum of the fifth dimension with electric charge. Using de Broglie's idea that wavelength is inversely related to momentum, he connected the maximum size of the extra dimension with its minimal amount of momentum, and linked the latter, in turn, with the minimum electric charge. The tiny magnitude of an electron's charge, he found, led naturally to a minuscule size for the fifth dimension. Consequently, the fifth dimension would be far too small to detect.

The undetectability of Klein's fifth dimension is like standing on a high ladder and observing a needle on the ground that is wrapped tightly with thread. From that high vantage point, the thread's thickness wouldn't be apparent and the needle would seem like a simple straight line. Similarly, because the fifth dimension would be coiled so snugly, it would be unobservable.

After completing his work, Klein was shocked to hear from Pauli about Kaluza's similar idea for unification. Pauli was one of the few who could keep track of all the developments and theories in general relativity and quantum physics, and he served as a font of information for others. Despite his disappointment at not being the first to consider five-dimensional unification, Klein decided that his theory was unique enough to publish. In later unification models, including some of Einstein's ventures, Klein's notion of a tiny, wrapped-up fifth dimension would prove to be a vital component. As a result, higher-dimensional schemes for uniting the natural forces are often called Kaluza-Klein theories.

Klein's approach had little impact on the Copenhagen community at the time, however. Bohr steered the group toward forging a consensus about the nature of the atom and of the quantum. That common ground included acceptance of the atom as a probabilistic mechanism. Neither Klein's five-dimensional idea nor Schrödinger's interpretation of waves as charge distributions included the notion of sudden quantum jumps, and so they were excluded from the emerging canonical view.

Schrödinger's October visit was thus like a seminary student of one faith speaking to an assembly of devout supporters of another and trying to defend his minority creed. Though his views often turned out to be fluid, the proud, stubborn Viennese physicist would not be quick to concede. He changed his mind on his own terms, not through cajoling.

Schrödinger arrived by train on the first of the month. After the long journey, he met Bohr at the station and was immediately bombarded with questions. The interrogation didn't cease until he gave his talk and started off on his trip home. Even when he developed a cold during his visit and was sick in bed, Bohr continued to probe his views. He was staying at Bohr's house, so he really had no choice.

Despite the barrage of queries, everyone in Copenhagen was friendly and gracious, especially Bohr's wife, Margrethe, who always made sure guests felt most welcome. Nestled in the warm, cozy household, Schrödinger came under intense pressure by Bohr, Heisenberg, and others for him to accept Born's interpretation and discard the idea of physical waves. Schrödinger resisted with all his intellectual strength. He didn't want his visionary theory to become simply an abacus the matrix supporters could use to carry out their calculations.

The crux of Schrödinger's rebuttal was to declare that random quantum jumps simply weren't physical. He argued for a continuous, deterministic explanation instead. That was somewhat of a turnabout, given that in his inaugural address for the Zurich appointment—echoing ideas expressed by his mentor Franz Exner—Schrödinger had emphasized the role of chance in nature and dismissed the need for causality in science. Schrödinger had also written to Bohr applauding how a theory of radiation he had helped develop, called the BKS (Bohr-Kramers-Slater) theory, had circumvented causality.[17]

Einstein had been vehemently opposed to the BKS theory, precisely because of its randomness. On that matter he and Schrödinger had been on opposite sides of the fence. But that was in 1924, before Schrödinger had his own causal, continuous, deterministic equation to defend. As chance would have it, by late 1926 mutual opposition to the notion of random quantum jumps forced the two of them into the same anti-Copenhagen camp. The alliance would be forged once they realized that they were among the few vocal critics of Born's reinterpretation of the wave equation.

After returning to Zurich from Copenhagen, Schrödinger continued to defend his disdain for quantum jumps on the basis that atomic physics should be visualizable and logically consistent. Bohr maintained hope that Schrödinger would come around to the consensus view, simply because wave mechanics, in its probabilistic form, meshed so well with matrix mechanics. At that point, quantum theory was still gelling, so divergent interpretations did not hinder its progress. A larger problem for Bohr's goal of achieving harmony was Einstein's far more vehement and vocal opposition.

Does God Play Dice?

By the end of 1926, Einstein had drawn a stark line of demarcation between himself and quantum theory. Irritated by the lack of attention paid to the notion of continuity, which he saw as a logical part of nature, he began to draw on religious imagery to make his case. Why religion? Einstein had grown up in a secular Jewish household and certainly wasn't devout. Nevertheless, he was often reminded of his Judaism in a negative way by the anti-Semitic attacks on his work by right-wing German nationalists and in a positive way by the movement for a Jewish homeland in Palestine, to which he had lent his support.

Despite Einstein's philosophical differences with Born, the two of them were close friends. They enjoyed intellectual discussions and playing chamber music together, and kept up a steady correspondence. Born came from a similar secular Jewish background. Given their commonalities, it is perhaps not surprising that Einstein appealed to Born, trying to convince him that quantum physics required deterministic equations, not probabilistic rules.

"Quantum mechanics yields much that is very worthy of regard," Einstein wrote to Born. "But an inner voice tells me that it is not yet the right track. The theory . . . hardly brings us closer to the Old One's secrets. I, in any case, am convinced that He does not play dice."[18]

As we have seen, the "Old One" was one of Einstein's shorthand expressions for God—not the God of the Bible, but rather the God of Spinoza. That was not the last time Einstein would make that point. For the rest of his life, in his explanations of why he didn't believe in quantum uncertainty, he would reiterate again and again, like a mantra, that God does not roll dice.

The quasi-religious tone of his statement was an appeal to reason and common sense, rather than a call to replace science with faith. He could well have said "My sense of natural order informs me that the laws of physics are not random," but he opted for the more dramatic statement. Indeed, the saying "God does not play dice" has reverberated in a way that "Natural laws are not random" would not have.

The dramatic quality of his pronouncement pointed to a growing confidence about the import of his statements. He had started to become used to his words being picked up by the press and broadcast to the public. Perhaps that is why, even in a private letter, he was theatrical in his plea.

As another way of trying to rebut Born's interpretation, on May 5, 1927, Einstein delivered a talk at the Prussian Academy purporting to prove that Schrödinger's wave equation implied definitive particle behavior, not just dice-rolling. The following week, he wrote to Born with a sense of triumph, "Last week I presented a short paper to the Academy in which I showed that one can ascribe fully determined motions to Schrödinger's wave mechanics without any statistical interpretation. Will appear soon."[19] Einstein submitted the paper to a prestigious journal. However, perhaps because he was unsure of his results, Einstein retracted it only a few days later, and it was never published. Only the first page of his aborted proof has been preserved for history.

Despite his prominence, Einstein's entreaties had little impact on the quantum faithful. Experiment after experiment showed that quantum mechanics was a highly accurate theory of atomic behavior. It matched prediction after prediction on the mark. Young researchers, unschooled in (or at least unmoved by) the philosophical considerations that motivated Einstein and Schrödinger, witnessed the empirical verifications and saw quantum mechanics as the only road forward. They were loath to argue with experimental success.

Unmoved by Einstein's arguments, Born continued to advocate his probabilistic interpretation. He cringed at the notion that everything in nature was predetermined. Why welcome a world without choice or chance?

Meanwhile, Heisenberg began to codify the indeterminacy in the quantum process of measurement with an influential paper that he sent to Pauli in February 1927 and published later that year, "On the Perceptible Content of Quantum Theoretical Kinematics and Mechanics." The title and theme of the article reflected Heisenberg's desire to combat Schrödinger's ardent call for "visualizability" with his own analysis of what could and could not be observed in nature.

Heisenberg's paper is notable for his introduction of what he dubbed the "indeterminacy principle," now usually called the "uncertainty principle," referring to the inability to measure certain pairs of observables simultaneously. Position and momentum form one such set; time and energy constitute another. In each pair, the more precisely you measure one quantity, the fuzzier the other. Although the mathematical reasoning behind this idea was developed earlier (the fact that order of operations matters for the matrices representing the paired quantities), it was in the 1927 paper that Heisenberg first tried to explain what was happening physically.

Heisenberg showed that if you wanted to measure the position of an electron, you would need to observe it with light. The minimum amount of light required would be a single photon. Yet that solitary photon, aimed at the electron, would collide with it, disturb it, and lend it extra momentum. Thus at the instant the electron's location was known, its momentum would be disrupted by an unknown amount.

Heisenberg also described the process that became known as "wavefunction collapse." Before a measurement is taken of any quantity, such as position, the wavefunction consists of a superposition (weighted sum) of eigenstates. As soon as the reading occurs, the wavefunction immediately transforms into one of the component eigenstates, ridding itself of all the other possibilities. Its position (or any other quantity) is then set to the particular eigenvalue corresponding to that eigenstate.

We can think of the collapse process by imagining a delicate house of cards positioned so that each of its sides faces a different compass direction. It teeters in a superposition of north, south, east, and west. Now picture a strong breeze coming along from a wholly random direction. By touching the structure, it is in some sense taking a measurement. The house of cards topples over in one of the directions, collapsing into one of its constituent eigenstates. The process of measurement has triggered a collapse from the superposition into a single position.

Hungarian mathematician John von Neumann would later show that all quantum processes obeyed one of two types of dynamics: the continuous, deterministic evolution governed by a wave equation (either the Schrödinger equation or a relativistic version such as the Dirac equation) and the discrete, probabilistic repositioning associated with wavefunction collapse. Schrödinger himself would continue to believe in the former process, arguing vehemently against the latter.

While largely an ally of Heisenberg in the battle for interpreting atomic processes, Bohr differed with him at first about the uncertainty principle. He didn't think it was useful to frame quantum philosophy by looking at measurement errors; rather, he thought that a deeper analysis was needed. He began to advocate a way of bringing all the different aspects of quantum theory together in a kind of yin-yang approach called "complementarity," which considered electrons and other subatomic objects to have both particle and wave properties, each of which is brought out through different kinds of measurements.

Bohr's complementarity considered the design of an observer's experiment. If a researcher was probing wave properties, such as interference patterns, he or she would plainly see such zebra stripes. On the other hand, if he or she were recording a particle property such as position, that quality would come into view through something like a dot on a screen. Such contradictions, Bohr came to believe, were a fundamental part of nature.

Soon, though, Bohr and Heisenberg agreed to present a united front about quantum measurement, with complementarity and uncertainty being alternative ways of looking at the same thing. Their combined views, including the idea of wavefunction collapse triggered by experimentation, eventually became known as the "Copenhagen interpretation of quantum mechanics."

Their unity was tested at the Fifth Solvay Conference on Electrons and Photons, held in Brussels in October 1927, when Einstein startled

Bohr and his supporters with his zealous antipathy to their views. Ehrenfest, who was friends with Bohr as well as Einstein, chided the father of relativity for being too closed-minded about another revolution in physics. He accused Einstein of opposing quantum mechanics the same way orthodox critics attacked relativity's novel aspects. However, Einstein would not cede ground.

The debates at the conference about quantum philosophy between Einstein and Bohr were largely informal, taking place mainly during breakfast rather than during the sessions themselves. Each morning Einstein would bring to the table a hypothetical situation in which quantum indeterminacy could be avoided. Bohr would think it over for a while, construct a careful rebuttal, and inform Einstein. The next day, the process would repeat. By the end, Bohr had successfully defended quantum theory against all of Einstein's objections.

Einstein returned to Berlin a far more isolated figure in the scientific community. While his world fame continued to grow, his reputation among the younger generation of physicists began to sour, as they derided his objections to quantum mechanics. With experimental findings continuing to support the unified quantum picture advocated by Bohr, Heisenberg, Born, Dirac, and others, Einstein's dismissal of their views seemed petty and illogical.

Schrödinger was one of the few who sympathized with Einstein's doubts. They kept up a conversation about ways to extend quantum mechanics to make it more complete. Einstein complained to him about the dogmatism of the mainstream quantum community. For example, he wrote to Schrödinger in May 1928, "The Heisenberg-Born tranquilizing philosophy—or religion?—is so deliberately contrived that, for the time being, it provides a gentle pillow for the true believer from which he cannot very easily be aroused. So let him lie there. But this religion has . . . damned little effect on me."[20]

In his retreat, Einstein endeavored to develop a unified field theory that would supersede quantum mechanics. Given the success of the quantum equations, few physicists were interested in Einstein's attempts. Einstein's papers soon became more widely covered in the press than noted by the physics community itself.

In retrospect, Einstein's contributions after Solvay made little impact upon science. They were largely mathematical exercises in exploring different possibilities for unification. Noting that Einstein developed no major theories after 1925, Pais quipped, "In the remaining 30 years of his life . . . his fame would be undiminished, if not enhanced, had he gone fishing instead."[21]

Although the physics community relocated to the realm of probabilistic quantum reality, leaving Einstein the lonely occupant of an isolated castle of determinism, the press still bathed him in glory. He was the wild-haired genius, the celebrity scientist, the miracle worker who had predicted the bending of starlight. He was something like a ceremonial king who had long lost his influence over the course of events; the media were more interested in him than in the lesser-known workers actually changing science. His every proclamation continued to be reported by the press, if largely ignored by his peers.

The perception that Einstein still had tricks up his sleeve persisted for the rest of his life. His unification theories developed in Berlin in the late 1920s helped keep him steadily in the public eye. Jilted by the mainstream physics community, who increasingly viewed him as a relic, he remained the darling of the international media.

CHAPTER FOUR

The Quest for Unification

This Einstein has proven a great comfort to us that
always knew we didn't know much. He has shown
us that the fellows that we thought was smart is just
as dumb as we are. . . . I think this Dutchman [sic] is
just having a quiet laugh at the world's expense.

—WILL ROGERS, "Will Rogers Takes a Look at the Einstein Theory"

Working in Berlin, Einstein was surrounded by constant activity. Not only was the city a major center for science and technology, it was also a haven for the arts. Unter den Linden, the main thoroughfare in central Berlin, offered in the late 1920s one of the most concentrated hubs of culture in the world. Stretching from the famous Brandenburg Gate to the central cathedral, city palace, and statue-packed Museum Island, it was home of the state library, state opera, and the main buildings of the University of Berlin.

Although inflation had racked Germany, Berlin had many bragging rights. The sprawling city boasted that it was the largest in area in the world. Pulsing new neighborhoods were popping up everywhere, packed with department stores, restaurants, jazz clubs, and other venues. Operetta companies were thriving, capturing from Vienna the title of best light opera scene. Bertolt Brecht and Kurt Weill skillfully mixed opera with street language and jazz in their masterpiece, *The Threepenny Opera,* which opened at the Theater am Schiffbauerdamm in August 1928.

Late in 1927, Planck retired from the University of Berlin. With the involvement of Einstein, Schrödinger was invited to fill the prestigious professorship. While Zurich held many attractions, particularly its proximity to the mountains, he was delighted to receive an offer. Returning to German soil once again, he and Anny happily moved to the bustling capital.

Anny recalled the excitement of those times: "Berlin was the most wonderful and absolutely unique atmosphere for all the scientists. They knew it all and they appreciated it all. . . . The theatre was at the height, the music was at the height and science with all the scientific institutes, the industry. And the most famous colloquium . . . My husband liked it very much indeed."[1]

Situated in the German capital, not only did Schrödinger become a central figure in the scientific community and have easy access to lectures and discussions, but he also started to enjoy a measure of international publicity. It was just a smidgen of the monumental attention awarded to Einstein, but it still gave him a taste of fame.

For example, in July 1928, *Scientific American* published an article that presented Schrödinger's view as being the canonical replacement for the Bohr model.[2] The *New York Times* took note and informed its readership that Schrödinger's theory was the new fashion. Bohr's work, it reported, was as out of style as "ankle-length skirts"; savvy readers would need to acquaint themselves with Schrödinger's wave theory of the atom instead.[3]

While Schrödinger began to enjoy publicity, Einstein started to detest it, except when it proved useful to the charitable causes he supported or earned him extra pocket money from popular articles and books that he published. Although Einstein felt that the public should be informed about science, he was dubious that many people could really understand his theories. Perhaps his bluntest expression of this was an unfortunate set of remarks that he made right after his 1921 visit to the United States in which he accused Americans of being boorish. His odd speculations about why they were interested in his work produced this headline in the *New York Times*: "Einstein Declares Women Rule Here. Scientist Says He Found American Men the Toy Dogs of the Other Sex. People Colossally Bored."

Einstein was quoted as suggesting that American women "do everything which is the vogue and now quite by chance they have thrown themselves on the Einstein fashion. . . . [I]t is the mysteriousness of what they cannot conceive which places them under a magic spell." American men, on the other hand, "take an interest in absolutely nothing at all."[4]

Elsa generally welcomed publicity and saw one of her roles as controlling and promoting Einstein's image. Nevertheless, as she discovered during a chilling encounter on January 31, 1925, major public figures often attract the unwelcome attention of deranged individuals. That day a Russian widow, Marie Dickson, forced her way into their apartment building. Brandishing a weapon—by some accounts a loaded revolver, by others a hatpin—she threatened Elsa and demanded to see the professor. Reportedly, Dickson was under the delusion that Einstein had been an agent of the czar. She had previous threatened the Soviet ambassador in France, served three weeks in prison, and then been deported. She had headed straight to Berlin to target Einstein.[5]

Knowing that her husband was up in his study, Elsa concocted a clever subterfuge. She pretended he was not at home and offered to call him. Dickson calmed down, left the house, and said she would come back later. Once she stepped out, Elsa phoned the police. Five police detectives arrived and were waiting for Dickson when she returned. After a violent struggle they arrested her and sent her to an asylum. All the while, Einstein was safely up in his study, immersed in his theories, not knowing until afterward that Elsa may have saved his life.[6]

Although Einstein may have been indebted to his wife, they often quarreled. His lack of interest in his appearance rattled her. He famously hated haircuts, which she had to persuade him to sit through, and also refused to wear socks. Given their elite status, she wanted him to look reasonable for photographers, but he couldn't have cared less. Maintaining a certain public image was just added pressure for him, and he preferred to be alone with his projects. He complained to her, in turn, about the "folly" of her expensive clothes.[7]

By the time the Schrödingers arrived in Berlin, stress—mixed with lack of exercise, overindulgence, and a heavy pipe-smoking habit—had begun to take a toll on Einstein's health. In March 1928, while visiting Switzerland, he collapsed and was diagnosed with an enlarged heart. After returning to Berlin, he was put on bed rest and a strict salt-free diet. He was incapacitated for many months, using that quiet time as an opportunity to work on a new unified field theory. That May, Einstein excitedly informed a friend, "In the tranquility of my sickness, I have laid a wonderful egg in the area of general relativity. Whether the bird that will hatch from it will be vital and long-lived only the gods know. So far I am blessing my sickness that has endowed me with it."[8]

Secrets of the Old One

Einstein's jubilee birthday year of 1929 was celebrated both publicly and privately. Publicly, it roughly coincided with the announcement of his first widely reported attempt at a unified field theory, spawned during his time of incapacitation. He had previously published other attempts at unification, to little fanfare. Turning fifty, producing new work, and being Einstein would award his novel approach ample press coverage.

Throughout the 1920s, other researchers' unification theories had whetted Einstein's appetite for unraveling the secret formula of the "Old One" that would describe how all the forces of nature meshed together. Gravitation and electromagnetism seemed to have too many similarities to be independent. Both were forces that weakened with the square of the distances between objects. General relativity's limitation was that it could accommodate only one of the forces, gravity. Its equations needed extra terms on the geometric side to make room for the other force. Adding additional factors to a successful theory was not a step to be taken lightly. There needed to be clear justification—if not through physical principles, then through mathematical reasoning.

Einstein had dabbled with variations of Kaluza's, Weyl's, and Eddington's ideas, but was not happy with the results. Try as he might, he couldn't find physically realistic solutions that resembled particles. He even produced a paper similar to Klein's five-dimensional theory only to realize that Klein had beaten him to the punch. Pauli had told Einstein about the similarity, prompting him to include an awkward note at the end acknowledging that its contents were identical to Klein's.

Then, starting in mid-1928 and persisting for several years, he turned to an idea called "distant parallelism" (also known as "teleparallelism" and "absolute parallelism"). His new approach juxtaposed Riemannian geometry with Euclidean geometry, making it possible to define parallel lines between two distant points in space. Starting with general relativity's curved, non-Euclidean spacetime manifold, he associated with each point an extra Euclidean geometry called a tetrad. Because the tetrads have a simple, boxlike, Cartesian coordinate system, Einstein noted that it would be very straightforward to see if lines within those structures are parallel or not. Such comparisons of distant parallel lines would add extra information that is not present in standard general relativity—allowing for the geometric description of electromagnetism along with gravity.

In standard general relativity, because of spacetime's curvature, each point has a differently oriented coordinate system—tilted differently from place to place. It is like looking at Earth from space. You wouldn't expect a rocket blasting off from Australia to head in the same direction as one launched from Sweden. Similarly, the directional arrows in the vicinity of one region of spacetime would be different from those in another. Consequently, in standard general relativity, one cannot determine whether distant lines are parallel or not. One can define distances between lines but not their relative directions.

Distant parallelism, with its boxlike additional structure, makes it possible to specify the relative directions of any two straight lines, along with the distance between them. It adds a navigation system for the universe that supplements the basic road map supplied by standard general relativity. For that reason, Einstein judged it more comprehensive.

Einstein's initial goal with each of his unified field theories was to reproduce Maxwell's equations of electromagnetism in a geometric way, bringing them under the umbrella of general relativity. He was pleased he could accomplish that with distant parallelism, at least in the case of empty space. He didn't, however, make testable experimental predictions, as he did with general relativity, or identify credible physical solutions.

He also didn't achieve his goal of reproducing the quantum rules. Starting in the late 1920s, for each of his unified proposals, he hoped that the equations would be overdetermined—meaning more equations than independent variables. Such redundancy, he hoped, would force the solutions to have discrete types of behavior, something like quantum levels.

An example of overdetermination would be writing down the equations for the motion of a baseball and adding an extra condition that its vertical position must have a certain height. While without the condition the baseball would have continuous motion, tracing a curved path through the air, including the condition would restrict its position to only two discrete values. It would reach that height once on the way up and once on the way down. Thus the continuous equations, in tandem, would produce discontinuous values. Similarly, Einstein hoped that an overdetermined unified field theory would force electrons into particular orbits, similar to the Bohr-Sommerfeld model and the eigenstates found via the Schrödinger equation. However, he couldn't achieve that goal.

In general, distant parallelism failed to reproduce either the classical or quantum behavior of particles, as much as Einstein tried to do so. Therefore, his proposal was largely a mathematical exercise rather than a rigorous physical theory.

Even the math for his theory wasn't novel. As Einstein belatedly learned, French mathematician Élie Cartan and Austrian mathematician Roland Weitzenböck had already published on the topic. Cartan reminded Einstein that they had once discussed distant parallelism at a 1922 seminar—an encounter that Einstein apparently had forgotten. Einstein would eventually give Cartan credit for the mathematics underlying his theory.

As it turns out, it is relatively easy to tweak general relativity to include a version of Maxwell's equations by tampering with its rules about lengths, directions, dimensions, and other parameters. Einstein thought at the time that distant parallelism offered a reasonable modification. His criteria included simplicity, logic, and mathematical elegance. However, as Pauli and others advised him, discarding general relativity's successful predictions, such as the bending of starlight, was a radical move that shouldn't be taken lightly. To his colleagues' dismay, Einstein's growing interest in abstract notions had pushed aside the need to match experimental data.

Walking on Air

In January 1929, Einstein prepared to release a short paper describing his new scheme for unification. Despite the lack of physical evidence, he issued a short press statement emphasizing its scientific importance and highlighting its superiority to standard general relativity.[9] As soon as the international press learned of the imminent publication, more than a hundred journalists clamored for an interview, hounding him for a simple description of his novel idea. Not realizing how abstract and unphysical the paper was, they sensed a breakthrough akin to relativity. Einstein refused at first to offer further comment, hiding from reporters.[10] Eventually he offered more detailed popular explanations, published in the *Times* of London, the *New York Times*, *Nature*, and elsewhere. The *Nature* piece quoted him as stating: "Now, but only now, we know that the force which moves electrons in their ellipses about the nuclei of atoms is the same force which moves our earth in its annual course about the sun, and it is the same force which brings to us the rays of light and heat which make life possible upon this planet."[11]

The announcement of the theory set off an avalanche of publicity, comparable perhaps to the 1919 eclipse announcement. Given its abstruse, hypothetical nature and lack of experimental verification, the amount of press it received was staggering. Almost a dozen articles referring to the theory were published in the *New York Times* alone.

Scientists around the world were asked to comment on and interpret Einstein's results. Enthusiasm abounded, despite the dearth of proof. Among the unjustifiably eager reactions was that of Professor H. H. Sheldon, the chair of New York University's physics department, who speculated wildly that "such things as keeping airplanes aloft without engines or material support, as stepping out of a window into the air without fear of falling, or of making a trip to the moon . . . are avenues of investigation suggested by this theory."[12]

The theory also seemed to strike a cultural chord. A number of clergy members remarked about its theological implications. One pastor, the Reverend Henry Howard of Fifth Avenue Presbyterian Church in New York, compared its message to St. Paul's preachings about nature's unity.[13] Humorists, such as lasso-twirling satirist Will Rogers, joked about its incomprehensibility.[14] Another jested that the theory could be used to test golf balls.[15]

Massive media attention to a theoretical physics article was virtually unheard of before Einstein. Einstein made even the most abstract, far-flung theory seem sexy, mysterious, and earth-shattering. The fact that his hypothesis offered a lifeless set of equations lacking experimental vital signs did not scare off coverage. Einstein's moving hand as he carefully composed his mathematical arrangements provided the press with all the vital evidence it needed.

Einstein cringed at his celebrity status. He clearly wanted the spotlight to be on his theories and their implications, not on him personally. Needless to say, the press focused on the physicist himself, to which his only recourse was to try, often unsuccessfully, to hide out.

In stark contrast to the booming public hype, the reaction of the theoretical physics community was barely audible. By that time, largely because of the quantum revolution, Einstein's ideas were rapidly losing relevance to the mainstream physics community. Among the most active quantum theorists of the younger generation, only Pauli kept up a keen interest in his work. While Einstein remained respected personally, his rapid-fire production of seemingly irrelevant unification proposals became seen as a joke. For instance, young physicists in Copenhagen mocked his ideas in a humorous production of *Faust* in which a king (Einstein) was besieged by fleas (unified field theories).

Pauli was not an easy audience to please. Keeping up his reputation for bluntness, he threw a sobering splash of ice-cold water Einstein's way. Commenting on an essay published about distant parallelism, he wrote in a letter to the editor: "It is indeed a courageous deed of the editors to accept an essay on a new field theory of Einstein for the '*Results in the Exact Sciences.*' His never-ending gift for invention, his persistent energy in the pursuit of a fixed aim in recent years surprise us with, on the average, one such theory per year. Psychologically interesting is that the author normally considers his actual theory for a while as the 'definite solution.' Hence . . . one could cry out: 'Einstein's new field theory is dead. Long live Einstein's new field theory!'"[16]

Privately, Pauli commented to Pascual Jordan that only American journalists would be gullible enough to accept Einstein's distant parallelism; not even American physicists, let alone European researchers, would be that naive. And Pauli bet Einstein that he would reverse course within a year.

Meanwhile, in contrast to the publicity granted to Einstein's results, little noticed at the time was Weyl's pivotal work at Göttingen showing that his old idea of gauge could be applied to electron wavefunctions and explain the electromagnetic interaction in a natural way. The reason is that including the extra gauge factor along with the electron's description mathematically requires the addition of a new "gauge field" that propagates through space. That extra field can be identified as the electromagnetic field, offering a gauge theory of electromagnetism. One can think of the gauge factor as a kind of fan that is free to point in any direction as it is spinning around. To keep it spinning requires the "wind" of an influx of electromagnetic field lines. Despite the brilliance of Weyl's quantum gauge theory of electromagnetism, it would take another two decades before the physics community began to make use of it. Pauli, who was very astute, would be one of the first to recognize its importance.

Rabbi Onion's Blessing for Unification

Once he got the word out to the public about his unification scheme, Einstein hastened to close the floodgates and push back the rising tide of paparazzi. His birthday was coming up soon, and he desperately needed to escape. Much to the confusion and dismay of the press, on March 12, two days before he would turn fifty, he elected to flee the official celebrations and hide out in a secret location. "Even his most intimate friends will not know his whereabouts," reported the *New York Times,* which pointed out that he had been "driven crazy" by questions about his unified field theory.[17]

Somehow, one anonymous reporter did locate Einstein's hideout and file a story about his private celebration. The savvy journalist found out that Einstein's wealthy friend Franz Lemm, known as "Berlin's shoe polish king," was lending out his villa in the woodsy district of Gatow for the occasion. Far from the glare of Berlin's city center, Einstein was celebrating his birthday quietly with his family.

When the reporter walked in, Einstein was peering intently through a gift microscope, gazing with wonder at a drop of blood that he had extracted from his own finger. Casually dressed in a floppy sweater, informal pants, and slippers, stopping on occasion to take a puff from a pipe, he exuded childlike contentment. Perhaps he recalled his childhood gift of a compass. Other presents he received included a silk gown, pipes, tobacco, and a sketch of a yacht friends were planning to have built for him.

Perhaps the most unusual tribute was a doll created by his stepdaughter, Margot, that depicted a rabbi holding an onion in each hand. Margot's passion was sculpture, and she specialized in mystical images of clerics. Shaping the rabbi figure for her dear stepfather was a labor of love. Proud of her work, she read him a poem about it, "Rabbi Onion."[18]

Rabbi Onion, Margot explained, was an extraordinary healer. Onions, according to traditional Jewish lore, are good for the heart. Einstein had attempted such a cure during his recovery the year before. She had fashioned the mystical sage with his magical onions to bless him with a long, healthy life. That way, he could compose many more unified field theories. Einstein winced at the thought of churning out more and more unification proposals—which turned out to be an accurate prediction.

When Einstein returned to his Berlin dwelling he found a mountain of gifts waiting for him. Foremost among his presents was a generous offer by the Berlin city government to obtain a house and land for him near the Havel River and the lakes it flows through so that he could enjoy the serene landscape and go sailing. The city offered him free use of a mansion on the Neu Cladow estate, recently procured from a wealthy gentleman. However, when Elsa arrived to inspect the residence, the former owner informed her that his sales agreement included the right for him to stay there indefinitely. Without mincing words, he asked her to get off the property.

Red-faced about the botched present, the city government scrambled to find a solution. After months of civic wrangling about the right plan for Einstein, the scientist decided to take matters into his own hands and buy his own property in Caputh, near Potsdam, right at the intersection of two lakes: Schwielow and Templin. He hired an ambitious young architect, Konrad Wachsmann, to design and build a comfortable wooden cottage for him and his family, just a short walk to forested trails and the lakes. During its construction his eagerly awaited sailboat arrived, called the *Tümmler* (porpoise). Once the house was finished and they moved in, he was truly in paradise.

By the Banks of Lake Schwielow

Caputh was a perfect place for Einstein to go hiking or sailing, which allowed him to escape into his thoughts and forget about the burgeoning demands on his time. In the sylvan retreat, he was as casual as possible, often going barefoot and either in pajamas or shirtless—never dressed formally. He deliberately didn't have a telephone, so those who visited him often stopped by unannounced. One time when a group of dignitaries was visiting, Elsa implored Albert to dress up. He refused, stating that if they wanted to see *him,* he was there, but if they had come to see his clothes, they were in the closet.

One of the frequent visitors to Einstein's cottage who didn't mind the casual atmosphere was Schrödinger, who similarly hated formal attire. While professors in German universities of that era were expected to wear a suit and tie to class, Schrödinger almost always wore a sweater. On sweltering days in summer, he would sometimes come in wearing just a short-sleeved shirt and pants. One time, a guard wouldn't even let him through the university gate because he looked so scruffy. A student had to rescue him by attesting that he really taught there.[19] In another incident, Dirac recalled that the hotel staff hesitated before letting Schrödinger into his fancy accommodations for the Solvay meeting because he looked like a backpacker.[20]

In July 1929, the Prussian Academy of Sciences honored Schrödinger by inducting him into its ranks. As the ceremony was a white-tie affair, Schrödinger dressed up. He gave a well-received talk about chance in physics, taking a balanced stance that did not either endorse or condemn the Heisenberg-Born view. He had learned to tread lightly around that touchy issue. In that way, he invited both the determinism and nondeterminism camps to use his equation the way they wished.

In general, Schrödinger was delighted to be part of such a prestigious organization as the Prussian Academy. However, he came to share with Einstein a feeling that the academy was rather stuffy. Both of them would much rather be hiking or sailing than suffering through dry meetings. Consequently, it was on the trails and waterways of Caputh that they truly bonded and became close friends.

In their walks in the woods and jaunts on the lake, Einstein and Schrödinger came to appreciate their common interests. Perhaps only Schrödinger's disdain for music prevented them from growing even closer; Einstein loved to perform chamber music with his dearest friends. At that point in the two men's lives, they shared a deep fascination with the philosophical ramifications of physics. Each was more at ease talking about how Spinoza's or Schopenhauer's views applied to modern science than about the latest experimental findings.

Einstein was much steadier in his opposition to the mainstream interpretation of quantum mechanics, though. Schrödinger's attitude was so changeable that in a talk held at a Munich museum in May 1930 he practically adopted the Heisenberg-Born interpretation of the wave equation, though he would retreat again several years later.

Einstein's firm position was expressed in a March 1931 interview in which he affirmed his belief in causality and opposition to indeterminacy. "I know very well," he stated wryly, "that my conception of causality as part of the nature of things will be interpreted as a sign of senility. I am convinced, however, that the concept of causality is instinct in matters related to natural science. . . . I believe that the

Schroedinger-Heisenberg theory is a great advance and am convinced that this formulation of the relationships of quanta is nearer the truth than any previous attempts. I feel, however, that the essentially statistical character of this theory will eventually disappear because this leads to unnatural descriptions."[21]

The clash between the two friends' views on cause and effect was featured in a news story published in the *Christian Science Monitor* in November 1931.[22] The piece was likely the first to mention both physicists' outlooks. Describing talks about quantum mechanics each delivered around that time, it compared Einstein's steadfast opinion that the law of causality still applied with Schrödinger's nuanced belief that physicists needed to become more open-minded about various alternatives, such as the prospect of acausality. Evolving perspectives, Schrödinger argued, might transform our way of looking at nature's behavior, including the possibility of rendering the law of causality obsolete.

We see that while both maintained an interest in philosophy, Einstein was more inclined to favor Spinoza's rigid view that the world's laws were set from the beginning and might be logically deduced, while Schrödinger favored a more malleable perspective, shaped by Eastern beliefs in the veil of illusion, in which society's changing viewpoint molds truth. What appears true today, Schrödinger argued, might be seen tomorrow as a misconception. Therefore, it is possible that we might never find the ultimate truth.

Along with their mutual interests in philosophy and its application to science, the two physicists had more mundane woes in common. Neither had a happy domestic situation; each had multiple affairs. Finding Elsa controlling, Albert sought ways to escape. She was dismayed when he attended concerts and theater performances with a stunning heiress, Toni Mendel, who flamboyantly rode around in a chauffeur-driven limousine. He also went out regularly with a blond Austrian beauty, Margarete Lebach, whom Elsa couldn't stand.[23]

Erwin and Anny had a strong friendship but little sexual spark, and they would never have children together. They had decided not to divorce, but rather to maintain an open marriage. They continued to find too much comfort in each other's company to split up completely.

In contrast to Einstein, who expressed regrets about his failures in married life, Schrödinger romanticized his various trysts and kept a diary of his exploits. Some of the affairs would last for many years. At one point he became smitten with a young woman whom he tutored in math, Ithi Junger. Their liaison led to her having an unplanned pregnancy. Though he strongly wanted a child, he wouldn't leave Anny. Against his wishes, Ithi had an abortion and left him.[24] While that affair was cooling down, Schrödinger began a relationship with Hildegunde "Hilde" March, the young wife of a physicist he knew from Innsbruck, Arthur March. Their passionate bond would end up becoming something like a second marriage.

Einstein and Schrödinger could not have realized how fragile and special their time together in Berlin and Caputh would turn out to be. The mirth, relaxed attitudes, and open-mindedness of those days would vanish without a trace once Nazi boots trampled the Weimar Republic. Accustomed to a cozy, celebrated life, both scientists would be forced into exile, never to sail the Havel lakes together again.

Ill Winds and Ocean Breezes

The early 1930s in Germany were marked by massive unemployment and unrest. The 1929 stock market crash set off a chain reaction that toppled a succession of teetering economies around the world, including the fragile German postwar engine. With the Nazi movement and other far-right groups stirring the pot of nationalism, German resentment of the armistice terms became a rallying cry for vengeance. Communists and socialists responded with calls for worker power, frightening many business owners and mainstream conservatives—some of whom came to see the Nazis as the lesser evil and a bulwark against communism. In Berlin, hundreds of thousands of unemployed laborers with nothing else to do were ripe for recruitment to political movements on both ends of the spectrum. A massive rally in Alexanderplatz, one of Berlin's main squares, was quashed by the police, using tanks to round up the demonstrators. The right and left battled for votes and supporters as weak coalition governments rose and fell.

While he wasn't active in any particular party, Einstein generally supported the progressive socialist movement and favored greater worker rights. He considered himself an internationalist, seeing nationalism as a dangerous force. As a pacifist, he supported the War Resisters' League. Generally straightforward about his views, he had no qualms about openly condemning the Nazis. While at first he saw the support for them as an aberration, he soon came to realize—even before they took power—the dire threat they posed. Schrödinger, in contrast, had no interest in politics and tended to avoid such discussions. He didn't take the Nazi movement seriously until it was too late.

During the economic crisis, both physicists worried about their finances and were open to opportunities to work abroad, at least temporarily. Einstein's chance came first. He was pleased to get a invitation to travel to Caltech, in Pasadena, California, in the winter of 1931 and visit Mount Wilson Observatory, where Hubble had discovered the expansion of the universe. The stipend of $7,000 that Einstein was promised for just two months was incredibly generous for the time— about a full professor's yearly pay.

By then, Einstein had the help of two paid assistants: Helen Dukas, his secretary, and Walther Mayer, his "calculator" (mathematical aide). Dukas handled Einstein's flood of correspondence and extensive calendar of speaking engagements. Mayer performed the routine mathematical manipulations required for Einstein's research, particularly in unified field theories. Einstein had started to realize that Pauli was right and that distant parallelism would not be physically viable. Therefore, he began to pursue other avenues for unification.

Before departing for the West Coast of the United States, Einstein published the *New York Times Magazine* commentary piece mentioned in Chapter Three, declaring his views on science and religion and advocating the Spinozan concept of a deity. The essay generated heated debate and helped focus public attention on Einstein's upcoming visit.

Massive crowds like those that would greet a visiting king or queen welcomed the arrival of Einstein and his entourage in the port of San Diego on December 30, 1930. His companions disembarking from the great ship *Belgenland* included his wife, Dukas, and Mayer. Elsa proved a critical translator; her English was far better than Albert's. Mayer was always on hand whenever Einstein had a free moment for calculations.

Once at Caltech, the physics faculty, led by famed experimentalist Robert Millikan, began to discuss with him the possibility of a permanent position. But given his attachment to Berlin—and the Caputh lifestyle in particular—the discussions were premature. Nevertheless, Einstein loved Southern California, especially Pasadena's beautiful gardens and mild weather. One highlight of his stay was meeting Hubble and seeing the Mount Wilson telescope. He and Elsa also took some time to hobnob with Hollywood stars such as Charlie Chaplin. A great fan of his movies, Einstein was honored to be his guest at the world premiere of *City Lights*.

The following winter, Einstein was invited to Caltech again for another two-month visit. The question of a permanent appointment resurfaced. Given all the problems in Germany and the frightening prospect of a Nazi-led government, Einstein was starting to consider emigrating. However, by then he had begun to receive other offers, including the possibility of an Oxford professorship.

In Millikan's wooing of Einstein, he made one fatal mistake. He introduced Einstein to educator Abraham Flexner, who had come to Caltech to discuss the establishment of an Institute for Advanced Study (IAS) in Princeton, funded by wealthy benefactors and dedicated to fundamental research. Flexner ended up recruiting Einstein for a position that was meant, at first, to be only part-time. He offered Einstein a whopping salary of $15,000 a year, which would make him one of the highest-paid physics professors in the country. Einstein insisted, as an added condition, that Mayer be established in a second permanent position for assistance with his unified field theory calculations. Flexner was stunned by Einstein's demand but eventually relented. Einstein, in turn, committed to the Institute appointment.

Around the same period, Einstein took the time to nominate Schrödinger and Heisenberg, in that order, for the Nobel Prize in Physics. As a Nobelist, Einstein had the privilege of suggesting candidates for that high honor. In his nomination he ranked Schrödinger first because, in his opinion, Schrödinger's findings were more far-reaching than Heisenberg's. Still, Einstein was generous to nominate Heisenberg at all, considering his opposition to Heisenberg's probabilistic views. He realized that many physicists placed the two of them on par, as co-founders of quantum mechanics. Therefore, he felt that it was logical to include both, with his personal preference duly noted.

In December 1932, the Einsteins and their companions set sail to Southern California for their third and final visit to Caltech. The visit was bittersweet, partly because Millikan was miffed about Einstein's new commitment and partly due to the growing realization that Adolf Hitler, who was then deputy chancellor in a coalition between a conservative party and the Nazis, was on the brink of leading Germany. As they stepped out the door of the Caputh cottage, Albert reportedly told Elsa it was the last time she would see it. Still, part of him must have thought there was a chance they would return, as he had written his colleagues in Berlin about plans there for the following year.

Ironically, Millikan had earlier booked Einstein to give a speech shortly after his arrival extolling German-American relations. The purpose was to court a donor. Not wanting to disappoint his host, Einstein delivered the speech, which he read in English from a translation of his own text. He used the opportunity to promote the idea of tolerance for opposing political views and religious beliefs, both in the United States and in Germany.

The mention of the United States alluded to public complaints by a right-wing group called the Woman Patriot Corporation that a known "revolutionary" such as Einstein was allowed into the country. Although nothing would come of it, the FBI began a file on him that for decades accumulated similar questions about his patriotism.

In stark contrast to Einstein's message of tolerance, about one week later, on January 30, 1933, Paul von Hindenburg, the president of Germany, appointed Hitler as chancellor. With a notorious racist and anti-Semite, backed by hundreds of thousands of brown-shirted paramilitary thugs called the Sturmabteilung (SA) or "storm troopers," grasping the reins of the German state, opponents braced themselves for caustic rhetoric at the very least. People wondered if Hitler would turn his hateful words into actions—or were they just political poses designed to attract bands of hooligan supporters?

Fire in the Reichstag

German politics was so changeable in the early 1930s that many pundits thought Hitler's chancellorship would be a passing phase. Moderate conservatives quietly expected that he would trump labor's support for the communists and move toward the center. As the economy improved, many thought that voters would come to their senses, elect more sensible politicians, and temper extremism. Even right after Hitler assumed his post, Einstein still harbored some hope of returning to Berlin. Schrödinger, though despising the Nazis and their intolerance, wasn't even concerned at first.

Then came a turn of events that no pundit had anticipated. On February 27, arsonists set fire to the Reichstag, the German parliament building. Although historians believe that the culprits were probably members of the SA, Hitler immediately pointed a finger at the communists. Parliament passed a law suspending civil rights and permitting indefinite detention of suspects. Communist politicians and other members of left-wing movements were summarily arrested and eventually sent to concentration camps. A new election was held on March 5, in which the Nazis became the largest parliamentary group.

Around the time of the Reichstag fire, Einstein came to realize that he couldn't return to Germany while the Nazis were in power. He wrote to Margarete Lebach that he had canceled a talk he was supposed to give to the Prussian Academy because he was afraid to set foot in the country. After leaving Pasadena by train and traveling to New York, newspaper reports that the Nazis had rummaged through his Caputh house horrified him further. In Manhattan, he gave speeches to various organizations decrying the Nazi assault on freedom. These were picked up by the German press, which slammed him for disloyalty.

In New York, Einstein and his entourage boarded the *Belgenland* for the return voyage to Europe. During the ocean journey, he wrote a polite letter to the Prussian Academy thanking them for their previous support but asking to withdraw his membership, citing the political situation as a reason to step down. Then, upon arrival at the port of Antwerp, Belgium, Einstein handed over his German passport to the consulate and renounced all of his ties to that country. For the second time in his life (the first being when he was a student in Switzerland), he was a man without a country.

Luckily, Einstein had many friends in Belgium and neighboring Holland who offered him assistance. Queen Elisabeth, who had been born in Bavaria and married into the Belgian royal family, was particularly supportive. Einstein held bank accounts in Leiden and New York that proved indispensable after the Nazis confiscated the money he had deposited in Berlin banks. Though homeless and stateless, he had a secure future abroad.

Einstein was fortunate to have left Germany in time. The Enabling Act, passed on March 23 by the German parliament, suspended all right to dissent, effectively granting Hitler complete power. The Nazis soon dissolved all provincial assemblies, solidifying their iron rule. The twelve-year dictatorship would be the most brutal the world has ever seen.

The Einsteins searched for a place to live temporarily until the IAS appointment was ready. They found a small house in Le Coq sur Mer, on the North Sea, to rent for the time being. The seaside cottage, though not as comfortable as Caputh, proved a cozy refuge for his months in Belgium, until he could leave for America.

It was a sad period for Einstein in many ways. Around the time he had he been forced to flee his native land, two of his dear ones met tragic fates. His son Eduard, nicknamed "Tete," who had done brilliantly in school and wanted to be a psychiatrist, began to suffer from schizophrenia and was committed to a mental institution in Zurich. Having corresponded with him about the world of psychology and the works of Sigmund Freud, Einstein had high hopes for his career and was devastated when it was cut short. Then in September 1933, Paul Ehrenfest, who had been one of Einstein's best friends, committed suicide. Before Ehrenfest killed himself, he had shot his own son Wassik, who had Down syndrome, with the delusional motive of sparing his wife the expense of caring for the child.

The cold, blue Atlantic would soon separate Einstein from Europe and its suffering. He would watch the situation from abroad, observing the lives of his former compatriots go from bad to worse. Never would he forget their plight, even when permanently exiled in the New World. Although he would never return to Europe, his pained heart and anguished thoughts would always remain there.

CHAPTER FIVE

Spooky Connections
and Zombie Cats

Cases might be quoted where the decision is really difficult,
serious, painful, bewildering, when we are down on our
knees to the Almighty to forgo it. But in this He is inexora-
ble! We must decide. One thing *must* happen, will happen,
life goes on. There is no [wave] function in life.

　　　　　—ERWIN SCHRÖDINGER, "Indeterminism and Free Will"

Schrödinger was a brilliant man but not a particularly brave one.
He yearned to be admired—by his peers, by the public, and by the
women in his life—and would often shape his words to help win over
his target. Never wanting politics or religion to serve as a barrier between him and others, he tried to stay as neutral as possible on sensitive
issues. While he did express philosophical views in his essays, these
were framed as intellectual ruminations, not as doctrine.

Nonetheless, the rise of the Nazis and their worship of male Teu-
tonic superiority was so antithetical to Schrödinger's character that
he could not possibly keep his feelings hidden. Unlike, for example,
Heisenberg, he disdained any form of nationalism. He loved foreign
languages, religious diversity, and exotic cultures. He saw no reason to
elevate the Germanic tradition and people above any others.

Anny recalled that Erwin's revulsion at Nazi practices once brought
him face-to-face with irate storm troopers. He strolled to Wertheim's,
one of the largest department stores in Berlin, only to discover that it
was being boycotted because of its Jewish ownership. The Nazis had
declared March 31, 1933, to be a national boycott of Jewish merchants.

Thugs with swastika armbands blocked customers from entering the store and picked fights with anyone they thought was Jewish. According to Anny, Erwin argued with the thugs, not realizing the danger, and was nearly beaten up. In the nick of time, young physicist Friedrich Möglich, a Nazi supporter, recognized him and intervened.[1]

Schrödinger had started to avoid the meetings of the Prussian Academy, perhaps sensing that it would become involved in the political situation. Indeed it did. On April 1, responding to Einstein's announcement that he was cutting ties with the organization and Germany in general, its leadership issued a harsh rebuke. In a widely publicized announcement, it openly condemned Einstein's "anti-German" behavior. Horrified by the action, Max von Laue, who was an active member, called for a vote to rescind the academy's statement. But none of the other leading members would stand up for Einstein—not even Planck, who had been a strong supporter. The vote failed and the statement was never withdrawn. Absent from the discussions, Schrödinger did not publicly take a stand.

Einstein would never forgive the academy's cowardly act. Aside from von Laue, Schrödinger, and to some extent Planck (who had expressed his support in private but not publicly), the academy members' abandonment of him was a bitter pill. The academy's refusal to defy the Nazis was one reason he would never set foot on German soil again, even after the war.

The academy's censure of Einstein was a tremor that signaled a much larger earthquake ahead. On April 7, the German parliament passed the heinous Law for the Restoration of the Career Civil Service, which barred Jews and political opponents from public positions, including teaching and academics. The only exceptions, at first, were veterans of the First World War who had served at the front, those who had lost relatives in the war, and those who had held their positions since before the war. Those exemptions would prove short-lived.

The university affected the most by the Nazi ban was Göttingen, which had many Jewish faculty members. Max Born, though one of the lions of quantum physics, was informed that he had to step down. Mathematicians Emmy Noether and Richard Courant similarly were dismissed. Nobel Prize–winning experimentalist James Franck resigned before he was asked to leave his position. Once again, von Laue tried to enlist support from his colleagues to condemn the purge, but to no avail. Planck, whose voice would have carried much weight, refused to protest the Nazi moves openly, though privately he was aghast at the developments.

Recruiters from universities in other countries soon realized that Germany's loss could well be their gain. The first to recognize the opportunity was Oxford physicist Frederick Lindemann, who set out to snare some notables to beef up his department's research. Thanks to J. J. Thomson, Ernest Rutherford, and others, Cambridge had leapt far ahead of Oxford in the sciences, and Lindemann hoped to make the situation at least somewhat more balanced. The haughty, posh, much-disliked Lindemann had set his eye on Einstein for a permanent position—but Einstein would commit only to brief yearly visits. The anti-Semitic law meant that others would likely follow Einstein's path out of Germany. Perhaps, Lindemann thought, they could be persuaded to make Oxford their new home.

Born in Germany and having studied at the University of Berlin, Lindemann was familiar with the country and followed its politics intently. Sensing immediately that the Nazi regime would pose a threat to the world, he shared his apprehension with Winston Churchill, one of his closest friends. During the Second World War, Churchill, as prime minister, would appoint him chief scientific advisor and help arrange for him to be admitted to the British peerage as Lord Cherwell. Lindemann would prove very influential in British military policy, famously (or infamously, depending on one's viewpoint) advocating the bombing of German working-class civilian housing. Ironically, given his future wartime role, around Easter 1933 Lindemann had little trouble riding in his chauffeur-driven Rolls Royce freely around Germany and meeting with a variety of academics.

At Sommerfeld's suggestion, Lindemann decided to pursue Fritz London, an accomplished quantum physicist who had developed key theories about how atoms bond into molecules. Visiting Schrödinger's house, the Oxford professor mentioned offering London a position. Much to Lindemann's surprise, Schrödinger asked that he be kept in mind if London decided not to accept. Lindemann hadn't considered the possibility that non-Jewish academics such as Schrödinger would consider leaving, but he agreed to broach the subject with potential funders of new Oxford positions.

A Call for Assistants

Schrödinger was well aware at that point of Einstein's success in procuring positions in other countries. Given his financial worries and animosity toward the Nazis, a post at Oxford sounded attractive. Like Einstein, however, Schrödinger made his acceptance contingent on hiring someone else to assist him. Schrödinger's equivalent of Mayer was Arthur March. He asked Lindemann if March could be given an Oxford appointment as well so they could work together.

There was a huge difference, though, between Einstein's and Schrödinger's motivations for wanting an assistant. After the age of fifty, Einstein had lost much patience with the nitty-gritty of calculations. Mayer was essential to his productivity. The situation with March was different. Schrödinger discussed the possibility of writing a book with him, but they never really collaborated. Rather, along with Arthur came his wife, Hilde, with whom Erwin was deeply smitten.

Lindemann returned to England and scampered to get funding for all the positions he had agreed to, including ones for Schrödinger and March. Meanwhile, conditions in Germany deteriorated even further. May was even worse than April. More dismissals of Jews took place. On Bebelplatz, right in front of the University of Berlin, a massive burning of books by Jews and other banned authors showed how much intellectual life had deteriorated. Born left for Italy, with the promise of a position at Cambridge.

In part to escape the mayhem, the Schrödingers and the Marches decided to spend their summer in Switzerland and Italy, including visits with Pauli, Born, and Weyl. Weyl had earlier been appointed to a position at Göttingen, but because his wife was Jewish, he had decided to step down and flee Germany. He would move on to a position at Princeton's Institute for Advanced Study.

In mountainous northern Italy, Erwin persuaded Hilde to take a long bicycling trip with him—just the two of them. During their excursion, their relationship became passionate. Sometime around then, Hilde got pregnant with Erwin's child. Rather than divorcing their spouses, they decided to work out an unusual relationship—essentially a complex marriage.

Lindemann met with Schrödinger again in September, in the beautiful village of Malcesine on the shore of Italy's Lake Garda.

He was excited to report that a British firm, Imperial Chemical Industries, had agreed to fund several positions, including a two-year post for Schrödinger and a separate visiting appointment for March. Schrödinger would be associated with Oxford's prestigious Magdalen College. Although the specific salaries were still being determined, Schrödinger had no desire to return to Berlin and warmly accepted. He, Anny, and Hilde moved to Oxford in early November. Arthur needed to negotiate a leave from Innsbruck, where he held a position, so he went back there for a time.

Schrödinger's departure from Germany angered the Nazis. He was the most senior non-Jewish physicist to leave. Heisenberg, though not a Nazi party member or supporter, was upset that he had abandoned Germany. In Heisenberg's view, loyalty to the German homeland and to German scientific progress transcended politics. One should wait out the regime and hope for a more reasonable government, he believed, not just take flight. To Heisenberg's credit, though, he strongly opposed the view, held by Philipp Lenard and Johannes Stark, that all "Jewish physics," such as the work of Einstein and Born, should be banned in favor of "German physics" (meaning physics by non-Jewish Germans). Heisenberg continued to maintain friendly contacts with Jewish physicists right up until the war began, and he resumed them afterward. He urged German Jewish physicists, such as Born, to try to stay as long as possible to keep scientific life active. In his eyes, therefore, Schrödinger's decision was a defeat for the German scientific community.

The Berlin that Schrödinger left bore little resemblance to the city he loved. Less than a year earlier the German capital had been full of life—artistic, scientific, political. Its avant-garde theater and operettas attracted international attention. It welcomed people of all faiths and viewpoints. By late 1933, however, it had become a cultural wasteland, open only to regime-sanctioned art, music, and theater. Discussing Einstein's contributions to theoretical physics had become taboo. The press was so regulated that only one newspaper noted Schrödinger's exit.

Soon came news that threw more sand in the Nazis' face while pumping up Lindemann's already inflated ego. Shortly after Schrödinger arrived in Oxford he learned that he had been awarded the 1933 Nobel Prize in Physics for his wave equation. The award would be shared with Dirac. Lindemann paraded his trophy around Oxford, while asking Imperial Chemical to bump up his stipend.

All was well until months later, when Hilde gave birth to Erwin's daughter, whom they named Ruth. Oxford was abuzz with the news that money had been channeled to allow one of its members to have a mistress. There was little hope, from that point on, for Schrödinger to obtain a permanent position at Oxford—even with his shiny new Nobel.

Subtle but Not Malicious

After spending much of 1933 in Belgium under the protection of its royal family, Einstein needed to bid farewell to Europe—forever, as it turned out. Albert, Elsa, Helen Dukas, and Walther Mayer set sail on the *Belgenland* one final time, arriving in New York on October 17. No crowds or reporters greeted his arrival this time. To avoid potential sabotage by Nazi spies, after they disembarked from the ship Einstein and his entourage were whisked by small boat to New Jersey and taken directly to Princeton.

As designated buildings for the Institute for Advanced Study had yet to be built, Einstein and other members shared space with the mathematics department in Princeton University's Fine Hall. One comfortable feature of the building was a seminar room with a large fireplace. Above the mantle was carved an expression of Einstein's in its original German. Translated, it read, "The Lord is subtle but not malicious." Einstein was expressing the hope that God would not mislead researchers to believe in a false theory of nature, even if finding the correct solution was challenging. Einstein still hoped to discover the ultimate theory that would unite all the forces.

One pressing issue Einstein faced was to find help with his calculations. Though he had gotten Mayer hired for precisely that purpose, his "calculator" decided to pursue his own mathematical research, much to Einstein's disappointment. To make matters worse, because Mayer's position was permanent, Flexner refused to provide Einstein with another assistant.

Because of Flexner's paranoid need to micromanage Einstein's schedule and keep him focused on his duties to IAS, the two of them soon became enmeshed in conflict. Einstein was mortified to discover that Flexner was censoring his mail and turning down invitations without consulting him. Flexner even turned down an invitation for Einstein to meet with the Roosevelts at the White House, although eventually

Einstein got word and accepted. Einstein felt like a prisoner, trapped at IAS with seemingly no one to help him with his calculations.

Fortunately, IAS hosted a steady stream of brilliant young researchers, eager to make their mark and work with established scientists. Two such youthful minds, Russian-born physicist Boris Podolsky, who had recently completed his PhD at Caltech, and American physicist Nathan Rosen, who had studied at MIT, were ripe for productive theoretical work. Einstein seized the opportunity, and they began to collaborate in a critical examination of quantum physics.

Despite his dislike for Flexner, Einstein was well aware of the dangers of returning to Europe. He realized that IAS, with its quiet location and freedom from teaching, offered him the best opportunity to pursue a unified field theory, tie up the loose ends of general relativity, and engage in other research close to his heart. So he decided to remain indefinitely.

One nice feature about Princeton was its relative proximity to beach locales where Einstein could go sailing. He bought a boat, which he named the *Tinef* (colloquial German and Yiddish for "junk"), and spent most summers in various communities on Long Island Sound and Saranac Lake in the Adirondack Mountains of upstate New York. As he didn't know how to swim, when occasionally his boat tipped over, he'd have to be rescued by local youth. That happened in the summer of 1935, when they were staying in Old Lyme, Connecticut, inspiring the *New York Times* headline "Relative Tide and Sand Bars Trap Einstein; He Runs His Sailboat Aground at Old Lyme."[2]

In another sailing incident, on Saranac Lake in 1941, a little boy may have saved his life when Einstein was underwater with his foot caught in netting. As the then ten-year-old rescuer, Don Duso, reported many years later, "He was down for the count. If I had not been nearby, he probably would have drowned."[3]

Knowing that they would likely reside in Princeton for a long time, he and Elsa began to look for a house. They found the perfect location only a few blocks from the university (and the temporary location of IAS), allowing him to walk or bicycle to his office. The shingled house they bought in August 1935 was at 112 Mercer Street. The upper floor was converted into his study and lightened up with a new picture window that looked out on the trees. Downstairs, the rooms were furnished with antiques they managed to have sent over from their former Berlin abode. He soon wrote to Queen Elisabeth of Belgium that although he felt alienated from society life, "Princeton is a wonderful

Albert Einstein's Mercer Street residence in Princeton, New Jersey. *Photo by Paul Halpern*

little spot. . . . I have been able to create for myself an atmosphere conducive to study and free from distraction."[4]

To make the place even cozier, they acquired a terrier named Chico and several cats. Chico was the front line in the battle to guard his owners' privacy. As Einstein remarked, "The dog is intelligent. He feels sorry for me because I receive so much mail; that's why he tries to bite the postman."[5]

One correspondent Einstein was happy to open letters from, however, was Schrödinger. They maintained a warm exchange, growing even closer philosophically in their isolation from their native lands. Einstein also continued to correspond with Born, whose opinions he greatly valued despite their sharp differences about probabilistic quantum mechanics. He tried to get Flexner to invite both of them to IAS, but to no avail. Flexner had washed his hands of helping Einstein.

Take My Wives, Please

Schrödinger did have the opportunity to visit Princeton, but through the university's physics department rather than IAS. The chance arose because of an endowed faculty position called the Jones Professorship, established by brothers who were Princeton graduates and wanted to expand research opportunities in math and science at the university. The invitation dated back to October 1933, when a committee of the physics department met secretly to decide whom to award the professorship. The chair of the committee, Rudolf Ladenburg, was a German émigré atomic physicist who was very familiar with the work of Heisenberg and Schrödinger and keen to invite them over. They decided to make the full offer to Heisenberg but also use some of the funds to invite Schrödinger for a period of one to three months. Schrödinger accepted, but Heisenberg declined, citing the political situation in Germany as a reason not to venture abroad.

Taking a break from his Oxford post, Schrödinger visited Princeton in March and early April 1934. Over the years, he had developed an impressive, eloquent lecturing style with a great use of vivid analogies. His literary interests such as poetry and theater served him well to help bring difficult scientific concepts to life. His extensive knowledge of ancient history and philosophy enriched his discussions of contemporary issues. Plus, he was perfectly fluent in English—clear and resounding with virtually no trace of an Austrian accent. Einstein, in contrast, could at that point lecture in English only if he read haltingly from prepared remarks, and he had a thick south German accent. The department was pleased enough with Schrödinger to suggest to the dean of science, Luther Eisenhart, that he be appointed to the Jones Professorship as a full-time scholar.

After returning to Oxford, Schrödinger thought hard about Princeton's offer but decided in the end to turn it down. The great attraction of Princeton would be working and residing in the same town again as Einstein. He had hoped that Flexner, prodded by Einstein, would make an IAS bid as well, but that was not to be. Eyeing Einstein's high salary and generous benefit of not having to teach, Schrödinger aspired to something similar, and to his disappointment, Princeton's offer—though generous by any standard—didn't match up. In wanting a position akin to Einstein's, Schrödinger didn't realize how unusual Einstein's situation was. Einstein was earning about 50 percent more than what prestigious universities such as Princeton were paying their senior physics professors. Citing salary as the main reason, he wrote to Ladenburg in October with his regrets.

Aside from financial reasons for not moving to Princeton, Schrödinger had to deal with his unusual family situation. Given his love for Hilde and his hope of spending time with baby Ruth—the child he had always longed for—he certainly didn't want to live across the ocean from them. He wondered how Princeton society would react if he brought them over, along with Anny. Could he even be prosecuted for bigamy? Reportedly he had mentioned the situation to the president of Princeton, John Hibben, and was disappointed by Hibben's negative reaction to the idea of a family with "two wives" and shared child care.[6]

In a parallel universe, Schrödinger would have accepted the Princeton position, become even closer to Einstein, and spent the rest of his life in comfort and safety. Maybe he could have found a way for Hilde and Ruth to emigrate discreetly. Instead, he would choose to move back to Austria just before it was about to be invaded and annexed by the Nazis, a development that would put him in peril and force him to escape. But causality depends on the past, not the future, and he had incomplete data, so his usually sharp mind made a very poor calculation.

Spooky Connections

By 1935, many quantum theorists, satisfied that their basic vision was correct, had moved on to the study of the atomic nucleus. With quantum theory considered settled, nuclear theory was where the action was. That year Japanese physicist Hideki Yukawa proposed a model for how nucleons (protons and neutrons) interacted via other particles called mesons, in what eventually became known as the strong force. Yukawa's theory attempted to explain how atomic nuclei hold together. (We now know that gluons, not mesons, serve as the intermediaries.) A little more than a year earlier, Italian physicist Enrico Fermi had begun to map out a process called beta decay, the transformation of neutrons into protons through the emission of electrons and other particles. That interaction, which explains certain types of radioactivity, ultimately became incorporated into the theory of the weak force.

While Schrödinger was interested in these developments, Einstein essentially ignored them. He preferred to focus on scoring a medley for the duet of his youth, gravity and electromagnetism, rather than introducing untested instruments and making it a trio or quartet. Hence, by the mid-1930s, his attempted unified field theories could no longer be construed as "theories of everything"; rather, they combined some, but not all, of the natural forces.

Meanwhile, Einstein continued to be troubled by the mainstream quantum approach. His last encounter with Bohr had been at the 1930 Solvay conference, where they had wrangled about the uncertainty principle. As in the Solvay meeting of 1927, Einstein had proposed a thought experiment purporting to contradict quantum notions, which Bohr, after much pondering, had refuted.

Einstein's hypothetical device was a radiation-filled box, equipped with a timer, designed to release a photon at a precise moment. By weighing the box before and after the release, one could calculate the exact energy of the photon, he argued. Therefore, in contradiction to Heisenberg's uncertainty principle, one could determine both the release time and energy of the photon simultaneously.

However, as Bohr cleverly realized, Einstein had forgotten to include the effects of general relativity. Using Einstein's own theory against him, he rebutted that the process of weighing the box—on a spring scale, for example—would slightly shift its position in the gravitational field of earth. In general relativity, the time coordinate of an object in a gravitational field depends on where it is located. Therefore, the position shift would lead to a smearing of the time value, consistent with the uncertainty principle. With quantum logic vindicated, Bohr outwitted Einstein once again.

Five years later, Einstein certainly hadn't forgotten his debates with Bohr. In a series of discussions, he brought up some of his quantum quibbles with Podolsky and Rosen. By that point Einstein readily conceded that quantum mechanics accurately matched experimental results about particles and atoms. However, as he pointed out to the young researchers, it could not be a *complete* description of physical reality. The reason was that if paired quantities such as position and momentum were real descriptions of nature, in principle they should have definite values for all times. Lack of knowledge of such values meant that quantum mechanics was not a comprehensive model of nature. Alternatively, if when position was measured momentum actually became fuzzy and unknowable, that would mean that it somehow blinked out of reality. Therefore, according to Einstein, the haziness of the uncertainty principle pointed to a limitation of quantum mechanics in matching theory to reality.

Another issue Einstein brought up was nonlocality, or "spooky action at a distance." Any remote, instantaneous influence from one particle to another would violate what he called the "separation principle." Causality, he argued, was a local process involving interactions between adjacent entities, spreading through space from one point to the next at the speed of light or slower. Distant things must be treated as physically distinct, not as a linked system. Otherwise a kind of "telepathy" could exist between an electron on Earth and one on, say, Mars. How could each immediately "know" what the other is doing?

By then, John von Neumann had formalized the notion of wavefunction collapse, originally suggested by Heisenberg. In that formalism, a particle's wavefunction can be expressed in terms of either position eigenstates or momentum eigenstates, but not both at once. It is something like slicing an egg. You could slice it along its length or across its width, but unless you want it to be diced instead of sliced, you'd only do one or the other. Similarly, when you "slice" the wavefunction of a particle, you are forced to choose between position and momentum components, depending on which of those factors you are trying to measure. Then, upon measuring the position or momentum, the wavefunction instantly collapses with a certain probability into one of its position or momentum eigenstates. Now suppose the cause for such a collapse is remote. The researcher, without giving the particle warning, decides which quantity to measure. How does the wavefunction instantly and remotely know which set of eigenstates it should be choosing from in its collapse?

The paper that resulted from the dialogue between Einstein, Podolsky, and Rosen, "Can Quantum Mechanical Description of Physical Reality Be Considered Complete?" (commonly called the EPR paper), was written and submitted for publication exclusively by Podolsky. Published in *Physical Review* on May 15, 1935, it created quite a stir among the quantum community—particularly with Bohr, who had thought that the debate had long been over. Bohr found himself having to defend quantum mechanics all over again, just when he had begun delving into nuclear theory instead.

The paper described a situation of paired particles—such as a system of two electrons—that moved to different locations, such as after a collision. Even though they are separated, quantum mechanics informs us that a common wavefunction would describe the joint system. Schrödinger would dub such a situation "entanglement."

Suppose a researcher measured the first particle's position. The wavefunction for the entire system would collapse into one of its position eigenstates, instantly revealing knowledge of the second particle's location as well. If, in contrast, the first particle's momentum were recorded, the second particle's momentum would suddenly become apparent. Because the second particle could not possibly know in advance what the researcher was planning to do, it must have both eigenstates ready—position as well as momentum. With its position and momentum eigenstates both existing at once, the second particle would find itself in a situation prohibited by the uncertainty principle. Rather than a seamless garment, the paper suggests, quantum measurement theory is a patchwork of contradictions.

Schrödinger soon wrote to Einstein applauding the results. "I was very pleased that . . . you openly seized dogmatic quantum mechanics by the scruff of the neck, something we had already discussed so much in Berlin," he said.[7]

However, as philosophers of science Arthur Fine and Don Howard each have pointed out, Einstein was careful to distinguish his personal views from the arguments expressed in the EPR paper. Surprisingly enough for such an established figure, Einstein never reviewed the paper before it was submitted. Therefore, he had some qualms about the way Podolsky constructed his line of reasoning. As he replied to Schrödinger, "[The paper] was written by Podolsky after many discussions. But it didn't come out as well as I actually had wanted; the essence was buried by erudition."[8]

Einstein did not want the emphasis to be on the truth or falsity of the uncertainty principle. Rather, he wanted to stress the need for natural laws that offered local and complete descriptions of all physical quantities. Quantum mechanics, as championed by Heisenberg, von Neumann, and others, appeared to have nonlocal and ambiguous aspects that called out for a more comprehensive explanation.

"All physics describes 'reality'," he explained to Schrödinger. "But this description may be complete or incomplete."[9]

To elucidate his point, Einstein described to Schrödinger a situation in which a ball could be in one of two closed boxes. Taken at face value, probability theory would suggest that it is half in one and half in the other. However, it couldn't really be split between both; it must be in one or the other. A complete description would state unambiguously where the ball is at any given time.

Even before the paper was published, Einstein let the world know about his views. On May 4, 1935, the *New York Times* offered the jolting headline "Einstein Attacks Quantum Theory." The article explained Einstein's view that "while it is 'correct' it is not 'complete.'"[10]

Einstein's Gunpowder

We've seen how, time after time, Einstein helped shape Schrödinger's ideas and career, from his interests in theoretical physics to his development of the wave equation, from his Berlin appointment to being awarded the Nobel Prize. True, Schrödinger had a brilliant, original mind. As the public today well knows, he developed the clever thought experiment of a cat in a box. However, Einstein inspired that too.

Einstein's EPR experiment helped reignite Schrödinger's antipathy to certain "fuzzy" aspects of quantum measurement. Schrödinger found a new zeal to explore inconsistencies in the standard view. In return, Einstein found in Schrödinger an eager ear for his qualms.

"You're actually the only person with whom I really like grappling. . . . You look at things as required, inside and out," Einstein wrote to him on August 8.[11] Almost everyone else, he felt, was wedded to the new dogma without objectively considering its disturbing implications. No doubt Schrödinger was delighted to have become Einstein's chief confidant on matters quantum.

In the same letter, Einstein went on to describe a paradoxical situation involving gunpowder. Experience tells us that gunpowder, assuming it is combustible, either has already exploded or hasn't exploded yet. But as Einstein pointed out, by applying the Schrödinger equation to the wavefunction representing a heap of gunpowder, it could evolve into a form where it is a bizarre mixture of the two possibilities. It would be exploded and unexploded at the same time.[12]

Hence, in Einstein's conception, large, familiar systems, expressed in the language of quantum mechanics, could well become monstrous hybrids that combine contradictory truths into a logically inconsistent reality. Logical inconsistency, including self-contradictory statements, was the fuel for Austrian mathematician Kurt Gödel's assertion, published in 1931 and presented at a Princeton IAS talk in 1934, that Hilbert's mathematical system was incomplete. Similarly, Einstein asserted that quantum mechanics contained self-contradictions that would topple its methodology.

A Cat's Strange Tale

Based in part on Einstein's gunpowder idea, with a measure of Einstein's ball-in-the-box thought experiment thrown in, Schrödinger crafted his feline thought experiment in a manner designed to highlight the ambiguities of quantum measurement. He acknowledged his debt to Einstein in a letter dated August 19, in which he announced that he had developed a quantum paradox that "resembles your exploding powder keg."

As Schrödinger described the imaginary experiment to Einstein: "A Geiger counter and a tiny amount of uranium that could trigger it are enclosed in a steel chamber—such a small quantity that within an hour there is equal likelihood that the counter would or wouldn't register a nuclear decay. An amplifying relay ensures that if atomic decay occurs a flask containing [poisonous] hydrocyanic acid would be smashed. Cruelly enough, a cat is also included in the steel chamber. After an hour, in the system's combined psi-function, equal parts of the living and dead cat would be blended—pardon the word."[13]

The implication is that before the box is opened and its contents revealed, just as the uranium has equal chances of having decayed or not, the cat would have equal chances of having been poisoned or spared. Therefore, the combined wavefunction representing both the Geiger counter reading and the cat's state would be in a strange juxtaposition—half decayed, half not; half dead, half alive. Only once someone opened the box would the combined wavefunction collapse into one of the two possibilities.

By positing a cat whose wavefunction is an equal mixture of life and death until an experimenter opens the box it is in, Schrödinger highlighted a situation even more implausible than Einstein's gunpowder scenario, hoping to show that quantum mechanics had become a kind of farce. Why a cat? Schrödinger enjoyed creating analogies involving familiar things, such as household objects or pets, to draw out the absurdity of situations by making them more tangible. It was not that he had a grudge against any particular feline—on the contrary, as Ruth recalled, he "loved animals"—or that there was one he wanted to immortalize.[14]

Could any two things be in an entangled state, no matter how dissimilar or distant? Might the wavefunction formalism, originally applied to electrons on a minute scale, be used to characterize anything at all? The mere idea of linking the fates of living beings and particles, he suggested, was ludicrous. Quantum mechanics had strayed far from its original mission if it could be applied to breathing, purring creatures.

Einstein wrote back to Schrödinger, heartily expressing approval. "Your cat example shows that we agree completely with respect to assessing the character of present-day theory," he said. "A psi-function, in which both the living and the dead cat are included, just cannot be regarded as the description of a real state."[15]

Much to Bohr's dismay, Schrödinger, in tandem with Einstein, seemed to be mocking a successful theory without providing a more credible alternative. What about a unified theory that superseded quantum mechanics? In no way, shape, or form would Bohr consider Einstein's (and later Schrödinger's) quest for a unified field theory credible, as the models Einstein advanced weren't based on atomic data and didn't even reckon with nuclear forces. Nevertheless, Bohr was always polite and patient, even with his detractors.

Schrödinger's kitty conundrum was published in November 1935 as part of a paper, "On the Present Situation in Quantum Mechanics"— the same article in which he had coined the term "entanglement." As we discussed in the introduction, the thought experiment was scarcely known by the public until many decades later. At that point, only the physics community had the opportunity to laugh, shriek, or grumble at Schrödinger's bizarre hypothetical scenario.

One of the motifs of the cat paradox is the clash between what goes on at the microscopic and macroscopic levels. As Schrödinger described in his paper, uncertainty on an atomic scale becomes linked with fuzziness on a human scale. Because such macroscopic murkiness is never observed, microscopic indeterminacy similarly mustn't exist.[16]

Schrödinger argued that probabilistic quantum rules could not apply to living beings. He was troubled by some of his contemporaries' assertion that quantum dice rolling explained the choices made by sentient creatures. Unlike with particle behavior, he pointed out, one could not develop a probability chart for actions made by people.

In the paper "Indeterminism and Free Will," written in English and published in July 1936 in the prestigious journal *Nature*, Schrödinger addressed the differences between particle interactions and human decision making, refuting analogies made between them. "In my opinion the whole analogy is fallacious," he wrote, "because the plurality of possible actions . . . is a self-deception. Think of cases such as the following: you are sitting at a formal dinner, with important persons, terribly boring. *Could* you, all at once, jump on the table and trample down the glasses and dishes, just for fun? Perhaps you could: maybe you feel like it: at any rate you *cannot*."[17]

In other words, preset factors, such as manners and personality, determine what decisions people end up making. Such a concept of "free will" seems closely tied to Schopenhauer's notion that seemingly spontaneous actions are really inevitable. If you knew people's underlying motives and background, you could generally predict what they would do under certain circumstances. However, there wouldn't be a case, according to Schrödinger, where you would say that they'd have a 75 percent chance of doing one thing and a 25 percent chance of doing another. Rather, you'd either anticipate correctly what they would do or you would fail to predict it correctly, depending on how well you knew them and the situation.

Schrödinger ridiculed the idea that Heisenberg's methods could be used to calculate how often people do certain things. "If my smoking or not smoking a cigarette before breakfast (a very wicked thing!) were a matter of Heisenberg's uncertainty principle," he wrote, "the latter would stipulate between the two events a definite statistics . . . which I could invalidate by firmness. Or, secondly, if that is denied, why on earth do I feel responsible for what I do, since the frequency of my sinning is determined by Heisenberg's principle?"[18]

An Offer He Should Have Refused

No historian has developed an algorithm that could accurately explain Schrödinger's decisions—not using the uncertainty principle or any other method. At the end of 1935, he found out that his position at Oxford would be funded for only two more years before it expired. He would need to move on—but to where?

Meanwhile, Arthur March took Hilde and Ruth with him back to Austria. Hilde was depressed and needed treatment at a sanitarium. With the mother of his child gone, Erwin took on yet another lover, Hansi Bauer-Bohm, a Viennese Jewish photographer who was then living in England. Like Hilde, she was a married woman, but she was far more confident and assertive. After they had spent many months together, she let him know that she planned to move back to her native city. With one of his lovers in Austria and the other soon to return, perhaps the die was cast for him to venture back there too.

As chance, fate, or the mysterious mechanisms of academic decision making would have it, Schrödinger received a tempting joint offer from two Austrian universities: a professorship at the University of Graz coupled with an honorary professorship at the University of Vienna. His old friend from his student days, Hans Thirring, arranged the latter. The only other offer on the table was a professorship in Edinburgh, which he briefly considered until he learned how low the salary would be. So he accepted the Graz offer, and the Edinburgh position went to Born, the second choice.

In hindsight, moving to Austria right before the Anschluss (its annexation by Nazi Germany) was an incredibly foolish move—especially for someone who had already angered the Nazis by leaving a prominent position in Berlin. As Anny remarked, "Anybody who thought a bit about politics would have told, 'Don't go to Austria. It is already very much in danger.'"[19]

The Austria that Schrödinger returned to was far different from the one he had left a decade and a half earlier. Since March 1933 it had been under one-party fascist rule, governed by a nationalist movement that became known as the Patriotic Front. Similar in spirit to the Italian fascists under Benito Mussolini, the party suppressed the Social Democratic left as well as the Austrian Nazi right. Engelbert Dollfuss headed it at first, until in July 1934 Austrian Nazis assassinated him in an attempted coup. The plotters' goal was unification with the German Reich under Hitler. When the coup failed, Kurt Schuschnigg took over as chancellor. He resisted pressure for Austria to align itself with Hitler, advocating its continued independence. However, the Austrian Nazi movement continued to grow. Like the German Nazis, it organized angry unemployed workers and other supporters into a formidable paramilitary force. They were encouraged by statements Hitler (who was born in Austria) had made supporting a greater Reich that included all German-speaking people.

In July 1936, Schuschnigg signed an agreement with Hitler that, on the face of it, seemed to guarantee Austrian independence. Austria and Germany promised to respect each other's sovereignty and not to meddle in each other's internal affairs. In return, Schuschnigg promised to ensure that his foreign policy was appropriate for a "German state" and to let some Nazi-leaning politicians into his government. These seemingly innocuous clauses provided a Trojan horse for Hitler to include his supporters in the Austrian leadership and begin to exert pressure from within for subjugation.

Schrödinger started his Graz professorship in October of that year. Once again he tried to ignore politics, focusing on his research. He had become intrigued by recent proposals by Arthur Eddington for uniting quantum physics with general relativity and explaining uncertainty through cosmological arguments. Thus in the midst of Austria's turmoil, his gaze was fixed on his equations.

The Quantum and the Cosmos

Eddington's role in the late 1910s and early 1920s as a leading defender, interpreter, and tester of general relativity had won him much respect in the physics community. However, starting in the mid- to late 1920s, his research had become increasingly focused on explaining the properties of nature through mathematical relationships that connected the very large and very small. Although he was in many ways a visionary who was one of the first to blend particle physics with cosmology, many physicists dismissed his later theoretical work as numerology rather than science. For example, British astrophysicist Herbert Dingle referred to his work (along with other speculative theories) as the "pseudoscience of invertebrate cosmythology."[20]

On the other hand, Einstein and Schrödinger greatly respected Eddington's independent-minded thinking. Like them, he was certainly not one of the herd. While they did not agree with his prescriptions, they appreciated his clinical look at the ailments of quantum mechanics and how to make it better.

Two of the most important relationships in modern physics are Schrödinger's wave equation and Einstein's equation of general relativity. Strikingly, their domains are very different. While Schrödinger's equation describes the distribution and behavior of matter and energy throughout space and time, Einstein's equation shows how the fabric of space and time is itself molded by the distribution of matter and energy. One key distinction between the two equations, therefore, is that in Schrödinger's equation space and time are passive, while in

Einstein's they are active. Another is that, at least in the Copenhagen interpretation of quantum mechanics, the solutions of Schrödinger's equation, wavefunctions, bear only an indirect relationship to what is actually observed. As the cat paradox so starkly depicts, observed quantities manifest themselves after an experimenter takes a measurement and causes the wavefunction to collapse into one of its constituent eigenstates. Naturally, no experimenter is needed for general relativity to produce definitive values. Otherwise, who would have been the observer during the 13.8 billion years of cosmic evolution?

Reshaping the Schrödinger equation to match up with special relativity turned out to be fairly straightforward, as demonstrated by Dirac in 1928. The Dirac equation, designed to describe fermions—particles with half-integer spin—yields solutions called "spinors": similar to vectors, but with a different way of transforming when rotated through abstract space. The algebra for dealing with spinor solutions of the Dirac equation, involving the multiplication of objects called Pauli matrices, is a bit more complicated than for wavefunction solutions of the Schrödinger equation.

The Dirac equation leads to an astonishing prediction that electrons have counterparts with opposite charge but the same mass. Dirac thought these were "holes" in the energy sea of the universe left over when electrons emerged. Rather, they turned out to be actual particles called positrons: the antimatter versions of electrons. Carl Anderson first identified them in 1932 through a study of cosmic rays.

Compared to its reconciliation with special relativity, linking quantum mechanics with general relativity turned out to be a far more formidable problem. Throughout the 1930s, many physicists attempted unsuccessfully to merge the two. Even Einstein, who generally stayed away from quantum matters except to critique them or attempt to supplant them, tried his hand. In his final years in Berlin, 1932 to 1933, he and Mayer worked on a way of expressing general relativity using four-component mathematical objects related to spinors, called semivectors.

Part of Einstein's motivation was to construct a unified field theory that allowed for oppositely charged particles of different mass: protons as well as electrons. All of his earlier unified field theories, including the distant parallelism approach, could handle only particles of the same mass, electrons. To bring protons into the picture, he and Mayer hoped to generalize the Dirac equation so that it conformed to general relativity and also predicted particles of different mass. Unfortunately, like his previous unified approaches, Einstein's semivector method failed to produce physically reasonable results. Once he moved to Princeton, Mayer didn't work with him anymore and he decided to abandon the semivector approach. It would be another in his used-car lot of theories taken for a multiyear test drive, found to be a clunker, and then traded in for another.

Eddington was similarly intrigued by Dirac's equation and tantalized by its bridge between quantum physics and the four-dimensional realm of special relativity. Along with Heisenberg's uncertainty principle, which appeared the previous year, it motivated him to develop a fundamentally new vision of the universe from the top down. In his analysis he started with a few basic propositions, such that the universe is curved and finite—similar to Einstein's original model of the universe with a cosmological constant—and that all physical quantities are relative. To measure a physical quantity such as position or momentum, Eddington suggested, a researcher must compare it to the values for other reference points. This comparison, within the context of a spacetime warped by gravity, introduces a measure of fuzziness, leading to the uncertainty principle. Because it is harder to measure smaller things by relating their positions and momenta to those of other known objects, uncertainty is much greater at the atomic level than at the astronomical level. Therefore, quantum uncertainty is not a fundamental feature of nature but the result of human inability to measure everything in the universe with absolute precision.

Regarding wavefunctions as composites rather than fundamental, Eddington used general relativity, modified by his idea of relative physical quantities, to map out distributions of positions, momenta, and other quantities for collections of particles. Then he combined this data to construct wavefunctions and wave equations. His goal was to show that the laws of spacetime, when viewed through the foggy lens of human limitations in ascertaining positions and momenta, led to equations resembling those of quantum mechanics.

Eddington developed an estimate for Planck's constant, based on the number of particles in the universe, the curvature of the universe, and other quantities. He argued that the discreteness of quantum jumps pertained to the universe having a finite amount of space and a finite number of particles. Treating the universe as something like a black-body, he calculated the energy available for each of its constituents, and thereby attempted to match Planck's figure.

While Eddington could be clear and engaging in his writings, his calculations pertaining to his fundamental theory, as he would call his connection between quantum and cosmos, were rather opaque. Always interested in the big picture, Schrödinger became fascinated by Eddington's theory but couldn't follow the steps he had taken to his results. In June 1937, Schrödinger wrote to him for clarification about his calculation of Planck's constant. Eddington responded, but still not to Schrödinger's satisfaction.

Italy was closely allied with Austria at the time, so it was relatively easy to travel there. In the course of 1937, Schrödinger journeyed there several times. A visit in June took him to Rome, to accept the honor of becoming a member of the Pontifical Academy of Sciences. On another trip, in October, he ventured to Bologna to deliver a scholarly talk about Eddington's theory. Much to his dismay, he received tough questions from Bohr, Heisenberg, and Pauli, who were in the audience, about Eddington's calculations. He was stuck in the precarious position of defending a theory he didn't really understand.

Despite Schrödinger's misgivings about Eddington's theory, it became a springboard for him to try to work out his own theory of unification. Like Einstein and Eddington, he began to see merit in explaining the troubling aspects of quantum mechanics, such as indeterminacy, jumps between states, entanglement, and so forth, through a greater theory based on modifying general relativity.

In Another Dimension with Unifying Intention

While Schrödinger was wrestling with the nuances of Eddington's fundamental theory, Einstein returned to the higher-dimensional realm of Kaluza and Klein. Coming full circle, he decided to try again to make use of the extra space provided by a fifth dimension to extend general relativity to include the laws of electromagnetism along with gravitation. Unlike his previous efforts with Mayer, he decided to include a physical extra dimension rather than simply a mathematical one. Adding a fifth dimension beefed up the equations of general relativity with five more independent components. By including those extra terms, he hoped to be able to describe the full behavior of particles—electromagnetic along with gravitational, and quantum along with classical. To work out the gritty details of the new approach to unification, Einstein was fortunate to employ two able assistants. The first, German Jewish physicist Peter Bergmann, joined the Institute for Advanced Study in September 1936. He had received his PhD in Prague under Philipp Frank, who had succeeded Einstein at that university. The second, mathematical physicist Valentine "Valya" Bargmann, also German born but of Russian Jewish ancestry, started the following year. He had completed his doctoral studies under Pauli in Zurich. As German Jews, each had little future in Europe; hence the move to America, where Einstein welcomed them. Remarking on the curious similarity of their surnames, Helen Dukas nicknamed them "the Berg and the Barg."[21]

Aside from meeting with his assistants, Einstein no longer had many constraints on his time. He had become a widower in December 1936, when Elsa died after a long illness that involved problems with her kidneys and heart. Her daughter Ilse had succumbed to cancer two years earlier. Dukas, who lived with the Einsteins on Mercer Street, had taken over most of the household responsibilities. Margot, and later Maja (Albert's younger sister), also resided with them.

Einstein developed a daily working routine. Each morning at around eleven o'clock Bergmann and Bargmann would stop by his house. They would have an informal chat and plan out the day, including time for calculations and possibly an evening enriched with chamber music. Dukas would see the three men to the door, making sure Einstein was appropriately dressed for the weather.

Einstein, Bergmann, and Bargmann would stroll through the leafy neighborhood until they reached Einstein's office at the Institute for Advanced Study. Until 1939, their destination would be Fine Hall, room 109, on Princeton's campus; afterward it would be Fuld Hall, the new headquarters constructed on the former Olden Farm just outside of the town center. As they walked, they would recount any struggles or victories they had had with their research since the day before. Most people listening in on their conversation would have had no idea what they were talking about.

Once settled in his office, Einstein would carefully check over their latest results and probe them with questions. His Fuld Hall office was divided into two parts: a large room with a big blackboard and a little room with a small blackboard. The two boards served different purposes. The big one, marked "Erase," was used for fleeting calculations that often led nowhere, assorted scribbles and notes, and anything else

Fuld Hall, where Einstein's office was located, Institute for Advanced Study, Princeton, New Jersey. *Photo by Paul Halpern*

that they deemed of transient importance. The small board, marked "Do NOT Erase," served as the hallowed *tabula* where the "ultimate" equations would be written.[22] Effectively, "ultimate" usually meant that they would last for a few weeks or months before being replaced. Nevertheless, in case they happened to be right, the sign prevented them from being scrubbed away.

By then, Einstein's criteria for whether or not equations were correct had veered sharply away from the world of experience. Although he remained nonreligious in the conventional sense, his cosmic religion, based on Spinoza, guided his judgments. He frequently asked his assistants to think about what options God would have chosen in designing a theory of everything.[23] Singularities (points in which quantities become infinite) and any other values that could not be determined by the equations were "sins," as he would put it. The equations should be as tight as an architectural blueprint, leaving nothing to chance.

Given his desire for a complete description of the universe in which there were no loose ends, Einstein's newfound zeal for the fifth dimension was somewhat of a dodge. Use of such an extra dimension allowed for nonlocal connections between distant things, as long as such links resided in an unobservable higher-dimensional enclave. In the EPR experiment and in his correspondence, Einstein had argued vehemently against the wavefunction having hidden information about a particle. All physical quantities needed to be "real" at all times, even if they weren't being measured. Yet in his five-dimensional unification attempts, information could be buried in an inaccessible space. It was like a politician telling the press, "While my opponent does not document any ties with overseas corporations, I do so fully—in papers locked away forever in my inaccessible security vault."

The main advantage of five-dimensional unification was that general relativity itself could remain untouched. The additional dynamics could be constructed in such a way that the four-dimensional description of gravity, matching the eclipse measurements and other experimental tests, would be preserved. Some of Einstein's other unification proposals, such as distant parallelism, did not retain these important results, making them suspect from the start. The equations that Einstein hoped would supersede quantum mechanics would result from the additional terms that arose from extending the number of dimensions from four to five. It was like the owner of a stately historical mansion deciding to build an addition to accommodate her needs for extra space, rather than remodeling the existing structure and ruining its appeal.

Einstein's assistants admired his perseverance. They would steam ahead with an idea for unification, day after day, until reaching a barrier. Once he realized they had been on the wrong track, he would patiently steer them onto a new course, barely expressing frustration or regret. He had faith that they would eventually reach their goal; it was only a matter of time.

Futile Concessions

In the final months of 1937, Schrödinger's most pressing challenges involved juggling teaching duties, his research interests, and time with the three different women in his life: Anny, Hilde, and Hansi (who had moved back to Austria, as expected). He had a seemingly secure professorship at Graz and a nice visiting position in Vienna, which gave him an omnipresent reason to visit his beloved native city and his good friend Thirring.

All that fell apart in early 1938, when the Anschluss brought Austria under the iron grip of the Nazis. Due to Hitler's unbridled ambitions and Germany's massive military superiority over Austria at the time, perhaps the conquest was inevitable. Schuschnigg tried desperately to appease the dictator while maintaining Austrian independence. His efforts culminated in a February 12 meeting with Hitler in which he agreed to coordinate his domestic and foreign policies with Germany and permit complete freedom for the Austrian Nazis. Then he miscalculated and decided to hold a plebiscite for Austrian independence, scheduled for March 13. Hitler was furious and ordered an invasion. Anticipating defeat, Schuschnigg resigned on March 11. When Nazi troops marched in the next morning and transformed Austria into a province of the Reich, there was no reported resistance.

Schrödinger was known to be an opponent of the Nazis and a close friend of Einstein. Disliking politics, he generally didn't see the need to broadcast his views. In Graz, where the Nazis were popular, he kept quiet about his beliefs. However, several weeks before the Anschluss he had given a talk in Vienna about Eddington's work in which, in closing, he had decried attempts by nations to dominate others. Inferring immediately which dominant power he had been talking about, the audience enthusiastically applauded him.

Following their takeover, the Nazis quickly purged the universities of socialists, communists, pacifists, Austrian nationalists, and anyone who was in the political opposition. All Jews were dismissed from universities and other public positions. Thirring, an ardent pacifist, immediately lost his job. Schrödinger could, of course, see the writing on the wall.

Tired of being a vagabond researcher, Schrödinger decided that he would try his best to retain his professorships, no matter what it took. Because of Hansi's Jewish background, he cooled off his relationship with her. Naturally, she was upset by his callous move. He also went to the Nazi-appointed rector of the University of Graz, Hans Reichelt, and asked for advice. Reichelt suggested that he compose a letter stating his loyalty to the Reich and send it to the university senate. Fearing being dismissed, he agreed to do so.

Much to Schrödinger's subsequent embarrassment, his statement of support for the Anschluss was sent to newspapers throughout the Reich and was widely published on March 30. Foreign scientists soon learned about it through a report in *Nature*. Former colleagues were stunned to read his "confession," which sounded like he was a born-again Hitlerite. "I had misjudged up to the last the true will and . . . destiny of my country," Schrödinger wrote. "The voice of blood calls [former doubters] to their people and thereby to find their way back to Adolf Hitler."[24]

In April, hoping perhaps to score further loyalty points, Schrödinger headed back to Berlin for a conference honoring Planck's eightieth birthday. His participation offered the possibility of turning back the clock and restoring his place in the German physics community.

Nonetheless, Schrödinger's gestures of solidarity with the regime proved futile. When he returned to Graz he soon found out that he had been removed from his honorary position in Vienna. By August of that year, he had lost his Graz professorship as well. The Nazis did not view him as reliable enough to retain his status. His Faustian bargain with the Hitler regime only landed him back in the purgatory of no academic role.

So Long, Farewell, Auf Wiedersehen, Adieu

Hollywood's dramatic depiction in *The Sound of Music* of one family's flight from Austria took some liberties. While the fictional von Trapp musical family fled stealthily across the mountains to Switzerland, the real von Trapps quietly escaped the Nazi regime by taking advantage of connections with Italy. Georg von Trapp had Italian citizenship, enabling them to travel freely to that country by train, then onward to London, and finally to America, where they had a concert tour already planned.

Similarly, after Schrödinger lost his academic positions and decided it was time for Anny and him to leave their native land, Italy proved a convenient exit route. However, their escape proved much more harrowing than the von Trapp family's. For one thing, although he had received indirect word of a new job possibility, the terms were very vague. Also, because Austria was no longer an independent country, he lacked proper travel documents.

Schrödinger's savior was someone he hadn't even met before: Éamon "Dev" de Valera, the *taoiseach* (prime minister) of Ireland. Born in the United States to an Irish mother and Cuban father, he moved with his family to Limerick, Ireland, at the age of two. After studying mathematics at Royal University in Dublin, where he embraced the work of William Rowan Hamilton, he lectured at St. Patrick's College in Maynooth and elsewhere throughout Ireland. In 1916, a growing sense of how Irish culture was repressed led him to join the Irish Volunteers and participate in the Easter Rising, a rebellion against British rule in favor of a democratic Irish republic. He commanded the Third Battalion from a post at Boland's Mill, a large flour warehouse.

Greatly outnumbered and outgunned by British troops, the Irish Volunteers were forced to surrender. De Valera and the other leaders were captured and, in all but one case, executed. De Valera's life was spared, possibly because of his American birth or maybe because there was pressure to stop the executions. After a year of jail time, he would return to Ireland to lead the Sinn Féin party and help establish the conditions for Irish independence. Due to differences with Sinn Féin over negotiations with Britain, he would eventually found the Fianna Fáil party and become *taoiseach*.

As party leader, he wrote the 1937 Irish constitution almost singlehandedly, and set the country on a course of neutrality and separation from the United Kingdom. Trained as a mathematician, he was deeply disturbed by the deterioration of Hamilton's former research center, Dunsink Observatory, seeing it as a symbol of decline. Not only did he want to bring renewed glory to Ireland, but he also set out to make it a leading force in mathematics and science. To that aim, he decided to plan out a Dublin Institute for Advanced Studies (DIAS), modeled after IAS in Princeton. But who would be its equivalent to Einstein?

After learning about Schrödinger's dismissal from Vienna, Dev decided that he would be a perfect candidate for a leading professorship at the planned institute. Because it would be unwise to contact him directly and risk alerting the Nazis, Dev sent out feelers through a chain of contacts. He spoke with mathematician E. T. Whittaker of Edinburgh, who had been one of his instructors in Dublin. Whittaker passed the word along to his colleague Born. Born wrote to Richard Bär, a friend of the Schrödingers who lived in Zurich. Bär asked a Dutch friend to travel to Vienna to let them know. The friend didn't find them there, as they were in Graz at the time, and so he left word with Anny's mother. Finally Anny's mother mailed them a brief note about de Valera's offer. Erwin and Anny read the note three times and then tossed it into the fire.

Erwin knew that he had little choice but to accept the offer. In his heart, he still wanted a permanent position at Oxford, but he sensed that it wasn't in the cards because of the funding issues and Lindemann's animosity toward him. His "confession to Hitler" had made Lindemann even angrier. Anny drove to Constance, on the Swiss border, where she met with Bär and conveyed their interest in a Dublin position. Bär wrote back to Born, who let Whittaker know. Whittaker, in turn, relayed the good news to de Valera.

On September 14, Erwin and Anny made their escape from Graz. Fearing that a taxi driver might report them, Anny drove their luggage to the train station and then left her car at a garage, asking that it be washed. That was the last time she would see it. With only 10 marks in their pockets, they boarded the train for Rome.

Arriving in the Eternal City, Schrödinger wanted to write to de Valera, and also to Lindemann, letting them know his status. He wanted to accept de Valera's offer while asking Lindemann if he could stay in Oxford in the interim. Fermi, who was a professor at the University of Rome, advised Schrödinger that any letters he sent might be censored. As Schrödinger was a member of the Pontifical Academy, Vatican City seemed a safer choice. Surrounded by the beauty of the Vatican gardens, he composed and sent the letters, addressing de Valera's in care of the League of Nations in Geneva. De Valera was serving at the time as president of that international organization. Two days later, de Valera called and invited them to Geneva for discussions. He had the Irish consul issue them first-class tickets and a pound each for expenses.

The Schrödingers excitedly boarded the express train for Switzerland. At the border they had a big scare when a guard, holding a piece of paper with their names on it, asked them to get off the train, separate, and go through security checks. Anny was very nervous putting her handbag and other personal items through the X-ray machine while officers glared at her. Luckily, they were allowed back on the train and made it to Geneva, where Dev warmly greeted them. After staying with him for three days, discussing plans for the institute, they headed to England.

Arriving in Oxford, Schrödinger was greatly disappointed by Lindemann's cold reaction. Lindemann would not forgive him for his pro-Nazi statement. Schrödinger didn't help matters by asserting that it was nobody's business and that he had done what he had to do. Fortunately, he didn't have to depend on Lindemann's assistance, as he soon got an offer for a one-year position at the University of Ghent in Belgium.

Given that the Dublin institute was still in the planning stage with no opening date in sight, he seized the opportunity.

Waiting for the Institute

Schrödinger learned more about the *taoiseach*'s plans during a brief visit to Dublin on November 19. As envisioned, the institute would include a School of Theoretical Physics and a School of Celtic Studies. Schrödinger brought up his own concerns, including his interest in having Hilde and Ruth join Anny and him in Ireland.[25] The request was highly unusual, given that Hilde had a husband of her own.

De Valera did not object, as Schrödinger's request was the least of his worries. He needed to get the Dáil's (Irish parliament's) approval for the Institute, a process that would take many months of political wrangling. While the Schrödingers were in Ghent, Dev argued with members of parliament such as General Richard Mulcahy of the opposition Fine Gael party, who saw the institute as redundant when there were already fine universities in Ireland that needed more funding. Critics derided the idea of combining fields as dissimilar as theoretical physics and Celtic studies into a single institution; the only thing the subjects seemed to have in common was that the *taoiseach* was interested in both. Perhaps the physics division should simply be dropped, Mulcahy contended.

Dev argued back that each branch could bolster the other's reputation. Invoking the legacy of Hamilton, he asserted that international accomplishment in the sciences would bring renewed glory and respect to Ireland. Because Fianna Fáil was in the majority, he knew that he could get the bill passed eventually. His arguments were targeted at those on the fence to help speed up the process.

Hearing about the debates, particularly the idea of scrapping the physics division, made Schrödinger nervous, but Dev assured him that all would work out in the end. He just had to be patient. As there were no other good options, the physicist had to trust in Dev.

One beneficial outcome of the year Schrödinger spent in Ghent waiting for the Dublin position was that he had the chance to get to know Georges Lemaître, the Belgian theoretician and priest who first proposed the idea that the universe expanded from a highly dense state—what later became known as the Big Bang theory. Schrödinger was inspired to contribute a calculation showing how certain types of cosmic expansion would entail the production of matter and energy. His result anticipated the steady-state theory of cosmology, advanced by Fred Hoyle, Thomas Gold, and Hermann Bondi in the late 1940s, and also the modern concept that much of the stuff of the universe was produced during a primordial inflationary era.

In his time of frustration, Schrödinger's thoughts turned to religious and philosophical questions, along the lines of Spinoza, Schopenhauer, and the Vedanta. An unpublished manuscript that he would bring to Dublin showed how his ideas about the search for natural order had converged on a belief system resembling Einstein's cosmic religion. As Schrödinger wrote: "In the presentation of a scientific problem, the other player is the good Lord. He has not only set the problem but has also devised the rules of the game—but they are not completely known, half of them are left for you to discover or deduce."[26]

By September 1939, the position at Ghent had expired and it was time for Schrödinger to leave Belgium. Hilde and Ruth had joined Erwin and Anny at their residence, with Arthur remaining behind in Innsbruck. Several complications had arisen. For one thing, the Dublin institute still hadn't been approved. Also, upon the Nazi invasion of Poland, the Second World War had begun. Not only was Schrödinger unemployed once again, but from the Allied perspective he was technically a citizen of an enemy power. That was a big problem, because he would have to pass through Great Britain to get to Ireland. Luckily, several of his benefactors, including de Valera and a surprisingly helpful Lindemann, intervened to get him and his extended family the proper papers to cross Britain and reach Dublin. They arrived there on October 7.

It was not until June 1, 1940, that the Dáil finally passed the law establishing the Dublin Institute for Advanced Studies. Its governing board met for the first time in November of that year. By that point, the major reason for the delay was the war. While Schrödinger was waiting, an apologetic de Valera helped arrange for him a visiting professorship at the Royal Irish Academy and courses to teach at University College Dublin.

Meanwhile, Schrödinger and his family found a house at 26 Kincora Road in the quiet suburb of Clontarf. It was a gorgeous location near the Bay of Dublin. A keen cyclist, he liked that it was close enough to the city center for a pleasant bike ride.

According to Irish cultural historian Brian Fallon, "the setting up of the Dublin Institute for Advanced Studies in 1940 was a landmark of

Erwin Schrödinger's residence on Kincora Road in the Dublin suburb of Clontarf. *Photo by Joe Mehigan, courtesy of Ronan and Joe Mehigan*

its kind."[27] It was a milestone in what some have called the "Gaelic Renaissance." Who better to direct its School of Theoretical Physics than a Renaissance man such as Schrödinger? When the institute opened in its Merrion Square headquarters, no one, save perhaps Dev, could be happier.

CHAPTER SIX

Luck of the Irish

[Einstein and Eddington's theory] did not work, they gave
it up. Why should it work now? Is it the Irish climate?
Well, yes, or perhaps the very favourable climate of 64
Merrion Square, where one has time to *think*.

—ERWIN SCHRÖDINGER, "The Final Affine Field-Laws"

Never rely on an authority in science. Even the greatest
genius can be wrong—whether he has one or two Nobel
prizes, or none.

—ERWIN SCHRÖDINGER, "The Final Affine Field-Laws"

In the heart of Dublin lies a gracious, green enclave, cozily surrounded by rows of grand Georgian townhouses. Close to Trinity
College, government buildings, and museums, Merrion Square was
a natural and beautiful setting for the Dublin Institute for Advanced
Studies. As a peaceful haven for scholars, de Valera chose well in
establishing the two branches of the Institute there: the School of
Celtic Studies and the School of Theoretical Physics. Later, a School
of Cosmic Physics would be established on the opposite side of
the square.

For the first time in years, Schrödinger felt safe and accepted. He
had plenty of time to explore new interests, such as biology, which
would culminate in an influential book, *What Is Life?* Proud of his
brilliant recruit, Dev would take the whole national cabinet to many
of the lectures.

64-65 Merrion Square in Dublin, where
the School of Theoretical Physics and
Erwin Schrödinger's office was located.
*Photo by Joe Mehigan, courtesy of Ronan and
Joe Mehigan*

Grateful to his host country and for the attention bestowed on him
by de Valera, Schrödinger aspired to become an expert in all things
Irish. He became fascinated by Celtic design. Visitors to his house
would notice his intricately crafted sets of handmade miniature fur-
niture, for which he wove the fabric on an Irish loom. He tried to
learn Gaelic and kept a primer, *Aids to Irish Composition*, on his desk.
Although he was adept at other languages, he had much trouble with
Irish grammar and eventually gave up. Nonetheless, many of his Irish
colleagues appreciated the effort. Above all, he would delight Dubliners
by telling them how much he preferred their city to snooty Oxford.

Schrödinger was active at the institute and friendly with his colleagues. As he typically worked late into the night, he never liked getting up early in the morning. Nevertheless, he would usually bike to
DIAS in time to join his fellow researchers for morning tea and pleasant
conversation.[1]

There were many reasons for Schrödinger to feel comfortable and
secure. Living in a neutral country, he could avoid the horrors of war

and the danger of expressing politically sensitive views. Moreover, with the powerful de Valera as his mentor and protector, he was free to pursue his unusual lifestyle.

Not only was de Valera the *taoiseach,* he was also the founder and owner of a major national newspaper, the *Irish Press.* The stances taken by that paper were typically aligned with those of de Valera, who was on the editorial board. Much later, in a major scandal, it was revealed that he had set up the paper's charter in such a way that the bulk of its profits were channeled to him and his family, despite thousands of outside investors from the United States and Ireland. The corporation was set up so that almost all of the investors held bogus shares and never collected dividends, while Dev and his family held the real shares and siphoned off the wealth.[2]

Reporters at the *Press* coped with crowded conditions and an atmosphere of mayhem. They knew on some level that it was part of their job to make Dev and his friends look good. Perhaps for that reason, along with his natural brilliance and charm that impressed reporters, Schrödinger received frequent and flattering coverage, often on the first or second page of the paper.

Take, for example, several cushy pieces published in the *Irish Press* about Schrödinger's domestic life. In the November 1940 article "A Professor at Home," he was described as "the biggest name in mathematical physics that the world knows in these modern times." The reporter had assumed Schrödinger would be aloof, but "when the gently-spoken man behind the cheery voice emanating quietly from a whimsically humorous mouth opened the suburban Dublin door, I knew I was wrong. This was a very human individual."[3]

As famous as Einstein was, his sailing excursions would be newsworthy only if he had a mishap, such as in 1935 when his boat ran aground on the Connecticut coast. For the *Press,* in contrast, even the fact that Schrödinger went on vacation was of note. For instance, when he decided in August 1942 to go on a bicycling trip to Kerry, the *Press* dutifully ran a story.[4]

Another article, "The 'Atom Man' at Home: Dr. Erwin Schrödinger Takes a Day off," published in February 1946, detailed Schrödinger's domestic life with Anny, Hilde, and Ruth. Nothing in the article painted their situation as unusual. It quoted but didn't question Schrödinger's misleading description of Hilde and Ruth's reason for being in Ireland. Pointing to Ruth, who had just beaten him in a game of chess,

Schrödinger said, "She and her mother, Mrs. March, were with us in Belgium when the war broke out. We brought them with us."[5]

Ruth was generally content in Dublin and enjoyed the attention of three parental figures. One day a good friend asked her why she had two mothers but her "father" (meaning Arthur) wasn't there.[6] Ruth didn't know. For her it was completely normal. She was very attached to the family dog, Burschie (roughly "Laddie"), a collie from the Wicklow Mountains that they obtained as a puppy, and was unnerved when he howled along with the test sirens during the war. Apart from the sadness she felt when Burschie died, she later recalled her years in Ireland as "quite uneventful."[7]

With de Valera as his patron, clearly Schrödinger had no need to fear a scandal about his love life. If anything, he felt more enabled to woo other women. With Anny and Hilde taking care of Ruth and the household, he continued to have multiple affairs. That was all behind the scenes, as recorded in his diary. Publicly, he was a "Great Brain," as the *Press* called him.

Each working day, Schrödinger biked from his well-kept suburban home to his comfortable office and back again. He took frequent holidays and only had to give a few lectures a year. His thoughts were free to wander through the wonderland of theoretical physics. In the wartime years, when many of his colleagues were suffering, de Valera had ensured a cozy life for him.

With all his good fortune, there was an understood debt that he needed to pay. He was expected to put the institute, and Irish science in general, on the map. When Einstein was appointed to Berlin, he had expressed worry that he was a "prize hen" that could well lose the ability to "lay eggs."[8] Schrödinger faced similar pressure to perform, with the added factor that the country's leader was looking over his shoulder at all times. He was seen as Ireland's best hope for international prestige in physics—its "new Hamilton," its only resident Nobel laureate, and its closest equivalent to Einstein. The press played up that image, setting the bar impossibly high for him.

Part of Schrödinger's impetus to create was internal as well. Bored with routines, he liked to challenge himself and reinvent himself. He liked to be seen as a Renaissance man—maybe even the heir of the ancient Greek philosophers. His active mind raced from topic to topic, hoping to find pathways to novel intellectual adventures.

Some physicists, faced with the need to innovate, would try to collaborate. In the early 1940s, however, such possibilities were limited. While Schrödinger remained well known among the international physics community, most physicists were focused on the war effort. Theoretical physics had moved into new directions, such as nuclear physics and particle physics. Schrödinger's interests had diverged from the mainstream.

Somehow DIAS, though centrally located, remained isolated from other Irish scientific institutions for a time. As Einstein's former assistant Leopold Infeld observed during a 1949 visit, "The Institute, which draws students from around the world, has put the name of Ireland on scientific achievement. Yet its influence upon its own country, upon Irish intellectual life and universities, is small."[9]

Laughingstock

While Schrödinger was the golden boy of the *Irish Press,* one of its rival newspapers, the *Irish Times,* while respectful, was somewhat less effusive. The *Times* kept a critical eye on the de Valera administration and its policies. In its quest to be independent and open, it often had to reckon with government censorship and accusations of libel.

From the perspective of Dev's opponents, particularly those in the Fine Gael party, DIAS was a vanity project for a pretentious leader who fancied himself a peer of the world's best mathematicians, scientists, and language scholars. Consequently, the *Times* took that institution a bit less seriously than did the *Press.* Until the paper was silenced by threats of a libel suit, one columnist even poked fun at its faculty in Swiftian fashion, painting its research as senseless and farcical, similar to that of the lofty island of Laputa in *Gulliver's Travels.*

The controversial piece appeared in the *Times's* humor column, "Cruiskeen Lawn" (Dublin slang for a full jug of whiskey), in April 1942. Penned by quirky, imaginative writer Brian O'Nolan under the pseudonym "Myles na gCopaleen," the column offered an irreverent look at modern Irish life. With a keen amateur interest in science and philosophy, coupled with a fluency in Gaelic, O'Nolan kept his eye on reports coming out of both branches of DIAS. He noted with curiosity that Professor F. O'Rahilly of the School of Celtic Studies had advanced the notion that St. Patrick was an amalgam of two different people. That seemed rather odd to him—perhaps even blasphemous.

O'Nolan also recalled that Schrödinger had delivered a talk in 1939 to the Dublin University Metaphysical Society entitled "Some Thoughts on Causality." As usual, Schrödinger had hemmed and hawed on the question, leaving open the question of whether or not the universe is causal. Realizing by then that he had gone back and forth on the issue as often as a porch swing in a windstorm, he quoted from Spanish writer Miguel de Unamuno that "a man who succeeded in never contradicting himself was to be strongly suspected of never saying anything at all."[10]

At the end of the talk, the president of the society, the Reverend A. A. Luce, had thanked Schrödinger for leaving open the possibility of free will, and thus being a "modern Epicurus." O'Nolan, however, had a different interpretation, and misconstrued Schrödinger's uncertainty about causality as doubts over whether there was a "first cause." In other words, according to O'Nolan, he had opened up the door to agnosticism. Without a first cause, there was no need for God.

O'Nolan's scathing column about DIAS focused on the two talks as examples of embarrassingly heterodox lines of research unsuited to its mission. "The first fruit of this institute," he wrote, "is to show that there are two Saint Patricks and no God. The propagation of heresy and unbelief has nothing to do with polite learning, and unless we are careful this Institute of ours will make us the laughing stock of the world."

O'Nolan also called the institute "notorious," remarking, "Lord, what I would give for a chair in it . . . for doing 'work' that most people regard as recreation."[11]

While Schrödinger took O'Nolan's comments magnanimously, seeing their humor, the DIAS leadership was enraged. They pressed the *Irish Times* for an apology. Its editor duly complied and promised that O'Nolan would never again mention the institute in his column.

Schrödinger had not been O'Nolan's only scientific target. After Eddington gave a July 1942 colloquium at DIAS about unification, explaining that relativity was truly understood by few people, O'Nolan proposed in his column that the subject should be taught in Gaelic to Irish schoolchildren. Instead of "being illiterate in two languages," he joked, they could be "illiterate in four dimensions."[12]

O'Nolan was also a novelist, publishing his fictional works under another pseudonym, "Flann O'Brien." One of his best-known novels,

The Third Policeman, was written between 1939 and 1940, overlapping with the period in which Schrödinger had arrived in Ireland and delivered his causality talk. Throughout his lifetime, O'Nolan couldn't find a publisher; the work was first published posthumously in 1967.

An offscreen character in the novel, an unconventional scholar named de Selby, makes himself known to the reader through a series of footnotes. De Selby espouses strange theories about nature, including a curious explanation for nighttime that involved a building up of "black air" due to volcanic eruptions and coal burning.[13]

O'Nolan's mockery of foolish scientific thinking has drawn much analysis. It is likely that the de Selby character was at least partly inspired by the lofty views of someone else whose name starts with "de": the cerebral de Valera. However, it is possible that Einstein and Schrödinger were also influences, given their prominence at the time.

The Stamp of Hamilton

No mathematician was dearer to de Valera's heart than Hamilton. To the world, 1943 was a year of devastation. German and Soviet forces battled fiercely in the struggle for Stalingrad. Ghetto fighters valiantly fought Nazi troops in Warsaw. However, for de Valera, 1943 was the year to celebrate the one hundredth anniversary of the quaternion, a mathematical entity invented by Hamilton in Ireland.

Quaternions are four-component generalizations of complex numbers. With real and imaginary (square root of -1) components, complex numbers can be expressed as points in a two-dimensional plane. Hamilton wanted to find the equivalent for points in three-dimensional space. His eureka moment came when crossing the Brougham (Broome) Bridge in Dublin. He realized that he would need four components, not just three. The definition of quaternions popped into his head, and he immediately carved the equations into the side of the bridge.

Under the leadership of de Valera, the Irish government issued postage stamps commemorating Hamilton and his discovery. De Valera hosted a gala in November of that year, inviting the international community to join the celebration. Because of the war, though, few foreign scholars were able to attend.

Why was de Valera so obsessed with pure mathematics in the middle of a world war? Even though Ireland was neutral, its economy was battered. Like many places around the world, food was rationed, and many supplies were limited. Yet Dev had a curious persistence about his personal interests that baffled his critics.

Anglo-Irish nobleman Lord Granard once noted, after a meeting with de Valera, that he was "on the borderline between genius and insanity." That could be said, of course, of many people with exceptionally focused vision. Against all odds and despite his unusual goals, Dev would remain politically popular—much like a bookish but admired schoolteacher who always seems to be looking out for the best interest of his pupils.

The start of the quaternion centenary meant additional pressure for Schrödinger to fulfill expectations that he would be the new Hamilton and restore scientific prominence to Ireland. By bringing Eddington to DIAS, along with other notables such as Dirac—who hadn't been to the Emerald Isle before—he had begun to put the country on the map. He had also helped recruit accomplished physicist Walter Heitler as an assistant professor, increasing the brainpower of the School of Theoretical Physics. Nevertheless, he carried the burden of justifying statements such as the following comment in the *Press*: "The man doing the most to continue the Hamilton tradition here is Professor Erwin Schroedinger."[14]

Given that Einstein was the standard for genius, Schrödinger adopted the paradoxical strategy of flaunting his connections with the esteemed physicist while subtly downplaying his achievements. Just as Sommerfeld had read Einstein's letters out loud to his classes, Schrödinger managed to let his colleagues and the press know that he and Einstein had a steady correspondence. Sommerfeld and Schrödinger clearly had different motives, though. Sommerfeld thought Einstein's words would inspire his students. Schrödinger enjoyed the bragging rights that went with having the public know about his friendship with Einstein. "The letters that pass between these two sets of brains are speckled with mysterious algebraic formulae that have the bulge on Lana Turner," the *Press* noted, referring to the Hollywood actress.[15]

Despite his close ties with Einstein, Schrödinger put him down in another piece about Hamilton. Remarking on the commemoration, Schrödinger wrote: "The Hamiltonian principle has become the cornerstone of modern physics, the thing with which a physicist expects every physical phenomenon to be in conformity. When some time ago, Einstein broached the idea of a 'theory without Hamilton principle' it caused a sensation. . . . In point of fact, it proved to be a failure."[16]

In some ways, Schrödinger at that time maintained a quantum superposition of outlooks that combined Einstein's attitude toward Heisenberg with Heisenberg's attitude toward Einstein. Along with Einstein, he would attack the believers in probability, such as Heisenberg, as being out of touch with mundane experience. The cat paradigm is a good example of that kind of critique. However, when he thought Einstein wouldn't hear, he would imply that the aging physicist had lost his grip—which was precisely the sort of thing that Heisenberg would suggest. It would take four more years, though, until Einstein fully came to realize how much he had been manipulated.

A Hermit in Princeton

In the wartime years, Einstein was rather isolated. Even in Princeton, a relatively small community, few came to know him. With Elsa gone, he had little incentive to dress up or even to cut his hair. Along with his various assistants, he continued his otherwise solitary struggle for unification.

As he wrote to a friend from Berlin who had moved to Haifa in the British Mandate of Palestine (now Israel), Dr. Hans Muehsam, "I have become a lonely old fellow. A kind of patriarchal figure who is known chiefly because he does not wear socks and displayed on various occasions as an oddity. But in my work I am more fanatical than ever and I really entertain the hope that I have solved my old problems of the unity of the physical field. It is, however, like being in an airship in which one can cruise around the clouds but cannot see clearly how one can return to reality, i.e. to the Earth."[17]

One reflection of the attitude toward Einstein was a humorous ditty composed by the Princeton University senior class of 1939. It was a custom for Princeton students to joke about their professors. Even though Einstein was never part of the university faculty, they chanted the following verse:

> To all the boys that study math
> Albie Einstein shows the path
> And though he seldom takes the air
> We wish to gawd he'd cut his hair.[18]

Working at the Institute for Advanced Study's newly constructed headquarters in the colonial-style Fuld Hall, Einstein no longer needed to share space with the math faculty on the university campus. He also didn't need to deal with Flexner, who had stepped down as director and was replaced by the mild-mannered Frank Aydelotte. In its bucolic setting, surrounded by acres of forest with many wooded trails, Einstein could enjoy long, pleasant walks with his colleagues—such as newly appointed member Kurt Gödel—as well as visitors.

One such visitor was Bohr, whom Einstein hadn't seen in years. He stayed for two months in winter 1939. Rather than the banter that would be expected during a reunion of the two sparring buddies, there was an eerie silence that divided them. Each was immersed in his own thoughts. Einstein, along with Bergmann and Bargmann, was fervently trying to find a five-dimensional extension of general relativity with realistic physical solutions.

Bohr had far graver concerns. From Austrian physicist Otto Frisch he had just learned about the successful experiments of Otto Hahn and Fritz Strassmann in Berlin, involving the bombardment of uranium with neutrons. Frisch's aunt, nuclear physicist Lise Meitner, had worked with them on the project before escaping to Sweden because of her half-Jewish ancestry. After analyzing the results, she and Frisch had concluded that nuclear fission (splitting of the nucleus) had taken place. After Frisch passed on the word to Bohr, he was horrified by the thought that the Nazis might discover the secrets of an atomic bomb. Indeed, during the war, Heisenberg would be put in charge of a nuclear effort that included Hahn and others.

Although his mind was on more pressing issues, Bohr courteously attended a talk by Einstein about his latest unification efforts. With a glazed expression he sat and listened to the founder of general relativity espousing a supposed "theory of everything" that seemed to ignore all developments since that era. What had happened to spin? To neutrons? To nuclear forces? He might have nodded off, except for the clincher. At the end of the talk, Einstein gazed directly at Bohr and said that his goal was to replace quantum mechanics. Bohr glared back but didn't say a word.[19]

Several months after Bohr's visit, Einstein was called upon to address the bomb issue himself. In July 1939, during a sailing vacation in eastern Long Island, Hungarian physicists Leo Szilard and Eugene

Wigner arrived at his house with a dire warning. They were gravely concerned that the Nazis would start to procure uranium from the Belgian Congo and use it to construct a bomb. Szilard had calculated that a chain reaction was possible in which neutrons released by the nuclear fission of one type of uranium prod more and more nuclei to break apart, producing substantial amounts of destructive energy.

In August, Einstein drafted a warning letter to President Franklin Roosevelt; Szilard translated it into English, and Einstein signed and sent it. A little more than two years later Roosevelt would establish the Manhattan Project: a top-secret effort, headed scientifically by J. Robert Oppenheimer, to develop an atomic bomb. While Einstein was never cleared to take part in the main efforts of the Manhattan Project, he would be asked several times during the war to contribute his knowledge to military projects. Meanwhile, while the world was greatly divided, he continued to press forward with his schemes for cosmic unity.

After working on unifying the forces for almost two decades, the normally optimistic Einstein would occasionally sink into bouts of despair. For example, in an address to the American Scientific Congress in Washington, D.C., on May 15, 1940, he "confessed that the task appeared hopeless, all logical approaches to the universe ending in a blind alley."[20]

Even in such dark moments, Einstein still refused to accept that the world was ruled by chance. While "there was no doubt that the Heisenberg uncertainty principle was true," he said at the conference, "he cannot believe that we must accept the view that the laws of nature are analogous to a game of dice."

Such self-doubt would be fleeting. Like voyagers seeking the Northwest Passage, if one route became blocked, he would try to find alternative byways to explore. Music would soothe his spirit while he planned out possible new avenues of investigation. Then he would consult with his assistants and resume his efforts along a different course.

In 1941, Einstein, Bergmann, and Bargmann published their swan song, a paper about five-dimensional unification. Bergmann left IAS that year for a position at Black Mountain College. Eventually he would form an important general relativity research group at Syracuse University and develop his own quantum theories of gravity. Bargmann would become a mathematics professor at Princeton. Once again, Einstein would need to find new assistants.

The Scourge of God

Einstein's next collaboration pertaining to unification would be with his old friend and frequent critic Wolfgang Pauli. Pauli saw it as his sworn duty to his friends and foes alike to be as brutally honest as possible. He wore with honor the tag Ehrenfest had bestowed on him: *die Geissel Gottes* (the scourge of God). He even sometimes signed his letters that way. Einstein seemed to appreciate the careful readings Pauli gave of his papers, but always needed to brace himself for the ruthless critiques. In a way, Pauli had a place in his cosmic religion. For the "sin" of misreading "God's thoughts" about natural law, he faced the torment of Pauli's ridicule.

In 1940, the Institute for Advanced Study's School of Mathematics invited Pauli to be a temporary member. Documents show that they selected him instead of Schrödinger (also under consideration) because they perceived that he would be less of a risk. They assessed Schrödinger as "brilliant, but less steady than Pauli. Already in 1937, when we compared their relative merits we decided in favor of Pauli."[21]

Why did the IAS School of Mathematics deem Schrödinger "less steady"? Could someone there have known about his unusual family situation? Given the rumors that he had discussed the issue with Princeton University's president, perhaps word had somehow spread to its neighboring institution. Alternatively, maybe Schrödinger's publication record, which included philosophical pieces as well as calculation-based papers, was seen as more sporadic. At any rate, Pauli was favored.

Pauli was pleased to leave turbulent Europe during wartime and venture to a quieter part of the world. While Zurich, where he worked, seemed relatively safe for someone with Jewish relatives, its proximity to Hitler's Reich certainly wasn't ideal. Thus he came to spend the war in Princeton.

Einstein decided to take advantage of being under the same roof with Pauli at Fuld Hall and attempt to provide common quarters for the forces of nature as well. Extending the ideas he had developed with Bergmann and Bargmann, Einstein worked together with Pauli on a five-dimensional unification model. It was one of the few cases where he collaborated with a noted physicist rather than with an assistant.

Pauli's careful approach led them to the unmistakable conclusion that there were no physically realistic solutions for such models that were free of singularities (infinite terms). The only singularity-free solutions they could find were massless and electrically neutral, such as photons. However, one of the goals of unification was to describe the behavior of charged particles with mass, such as electrons.

In 1943, Einstein and Pauli published a joint paper noting the lack of credible solutions. While one "cannot help feeling that there is some truth in Kaluza's five-dimensional theory," they remarked, "its foundation is unsatisfactory."[22]

Einstein's higher-dimensional aspirations had reached a dead end. He decided to abandon the approach taken by Kaluza and Klein and focus instead on theories with the standard number of dimensions: three of space and one of time. Although others would take up Kaluza-Klein theory and try to make a go of it, Einstein believed that he had exhausted its possibilities. His "Do NOT Erase" blackboard needed to be scrubbed. Clearly, it was time to move on.

Affine Frenzy

Ironically, just about when Einstein's progress toward unification had reached an impasse, Schrödinger started to become an enthusiast. Inspired by three of the theorists he admired most—Einstein, Eddington, and Weyl—he decided to try his luck too. He perused some of their early papers on general relativity and unified field theory and began to devise his own approach.

Because they were relatively isolated in their respective institutes, it was natural for Schrödinger and Einstein to correspond about subjects of mutual interest. Starting in winter 1943, Schrödinger began to write to him on a regular basis about the possibility of extending general relativity to include other forces. On New Year's Eve, he sent Einstein the kind of holiday greetings that only a theoretical physicist could convey. It was a letter deriving the equations of general relativity using the Lagrangian method, based on Hamilton's least-action principle. In a postscript, Schrödinger suggested modifying the Lagrangian and examining the field equations produced.

As we've discussed, Hamilton developed the least-action principle and Lagrangian method as ways of describing motion by envisioning objects as taking the most efficient path among all possibilities. It is like pioneers crossing a mountain range in the quickest way by trying to minimize their hiking and climbing. If they take elevation and other factors into account, the straightest path on the map might not be the best. Similarly, the route a particle takes through space depends upon the terrain of potential energy. A quantitative map of such terrain is incorporated in the Lagrangian, which can be used to find the equations of motion.

As Hilbert demonstrated, Einstein's equations of general relativity can be reproduced using a Lagrangian that consists of the product of two scalar (invariant under transformations) quantities, one related to the metric tensor that relays how distances are measured and the other related to the Ricci tensor (connected with the Einstein tensor mentioned earlier) that describes the curvature. The metric and Ricci tensors can each be expressed in matrix form as 4-by-4 arrays. Each would have sixteen components, but because of symmetry, only ten of the components are independent (the other six are duplicates). In standard general relativity, the ten independent curvature components are linked to the ten independent components of the stress-energy tensor, representing matter and energy. In short, matter and energy cause spacetime to curve, with ten independent relationships being involved.

Curvature, however, is only one aspect of spacetime geometry. To understand the paths taken by objects through space, you need to know the metric tensor components that tell us how to determine the distance from one point to another. These components produce an altered version of the Pythagorean theorem for that particular region. As we described in an earlier analogy, the metric tensor is like weaving a canopy above the hot desert sand in regions where the sand dips and rises (the curvature) due to scattered rocks (the matter and energy). To construct that metric canopy one needs to build a kind of scaffolding that tells us how poles (the local coordinate axes) bend from point to point. The links for that scaffolding are the affine connections. In standard general relativity there are sixty-four affine connections, and symmetry restrictions deem only forty of these to be independent.

Such is Einstein's standard description of gravity. To incorporate additional components associated with electromagnetism requires modifying the equations through options such as more dimensions (which Schrödinger didn't seriously consider), an extra structure such as distant parallelism (also not considered by Schrödinger), or relaxing the symmetry requirements and making the affine connections fundamental.

Following a path taken by Eddington and briefly considered by Einstein in 1923, Schrödinger elected to drop the symmetry requirements and focus on the affine connections. He called his approach the general unitary theory. (The acronym "GUT" would later be used for grand unified theories, proposed ways of uniting the electroweak with the strong force.)

Before Einstein had time to respond to the New Year's Eve letter, Schrödinger had already composed the rudiments of his approach. He started with the most general possible set of affine connections and used these to construct the Ricci tensor and a more flexible type of Lagrangian. This flexibility opened up the door to including electromagnetic components. He also hoped to add in components of what he called the meson field (what we now call the strong force) but decided to reserve that for future research (which he would shortly start to undertake). Then he used certain mathematical properties to restrict the Lagrangian to a special case, ending up with something different from Hilbert's Lagrangian. His equations made the unusual prediction that magnetic fields (such as Earth's and the Sun's) would drop off more quickly with altitude than conventional theory suggests. The attenuation was due to a kind of "cosmological constant" for electromagnetism, similar to the term Einstein had introduced many years earlier for gravity. With the basic ideas of his theory sketched out, he felt ready to report his results to fellow scholars.

Life, the Universe, and Everything

On January 25, 1943, Schrödinger presented his general unitary theory to the Royal Irish Academy. His paper would be published about five months later in the academy's proceedings. He explained in the talk how he had picked up where Eddington and Einstein had left off. In a year honoring Hamilton, Schrödinger was happy to make use of the Irish mathematician's methods, offering further tribute to him.

The *Irish Press* trumpeted the story, emphasizing the "beyond Einstein" angle. With the headline "Forward from Einstein," on February 1 reporter Michael J. Lawlor offered this startling announcement: "A scientific theory of importance so profound as to be comparable with Einstein's famous theory of relativity which revolutionized the modern physicist's conception of the nature of the universe has been evolved by

Prof. Erwin Schroedinger. . . . Einstein, it has been said, opened up a new world to the mind of man. Prof. Schroedinger, basing his conclusions on the mighty structure of the general theory of relativity, has now taken another huge step forward, a step so big that it may be that in the time to come the new theory will play a part like Einstein's has in our day."[23]

The next day, the *Press* ran another story that included interviews with several Irish scientists about Schrödinger's work. Dr. A. J. McConnell of Trinity College, one of those contacted, applauded his efforts, especially given that these are "difficult times for an institute of pure science. His colleague, Prof. C. H. Rowe, described Schrödinger's achievement "as an outstanding event in the history of science in this country."[24]

Schrödinger was scheduled that month to give three public lectures about a wholly different topic, "What Is Life?" Although he had no training or research experience in biology, as a youth he had absorbed some of his father's fascination with the science of living organisms, and he wanted to share his insights about the subject.

When he arrived at the Physics Theatre of Trinity College for the first lecture, he found the hall so packed that crowds of people were turned away. He agreed to repeat the lecture a few days later for those who couldn't get in. Naturally, the *taoiseach*, his biggest fan, was prominently seated in the audience of hundreds. At the end, Schrödinger received an ovation.

Some of the key insights Schrödinger covered involved the relationship between the properties of atoms and the behavior of living things. Pointing out that most natural systems tend toward increasing entropy (disorder), he showed how life maintains itself as orderly through the absorption of energy, such as from the Sun. He also speculated that an aperiodic crystal (a nonrepeating arrangement of atoms) played a role in the development of life. Hence he was one of the first to suggest that life was encoded by a chemical sequence. A book based on Schrödinger's lectures would serve as a source of inspiration for biologists in the 1950s, such as James Watson and Francis Crick, as they developed the double-helix model of DNA.

The popular lectures drew the attention of *Time* magazine, which reported, "Schrödinger has a way with him. His soft, cheerful speech, his whimsical smile are engaging. And Dubliners are proud to have a Nobel prizewinner living among them."[25]

When the *Irish Press* first reported about Schrödinger's general unitary theory, it sent Einstein a copy of the article to gauge his reaction. In April, Einstein finally cabled a restrained, polite response. "Prof. Schroedinger's is a very cautious and critical mind," he wrote, "so that every physicist must be highly interested in his new attempt to solve this formidable problem. I am not able to say more at the moment."[26]

In the same piece, the *Press* asked Schrödinger for his reaction to Einstein's response. "Of course Professor Einstein could not say more until he had seen the full scientific paper," he replied.

The exchange in the *Press* was cordial enough not to disrupt their friendship as of yet. However, as Schrödinger's confidence about his theory continued to grow, his pronouncements about its superiority to Einstein's work became bolder and bolder.

The Tomb of Einstein's Hopes

At a meeting of the Royal Irish Academy on June 28, Schrödinger generated excitement once again by claiming that his general unitary theory had been verified through experimental evidence. Explaining that he had resurrected an idea that Einstein had abandoned twenty years earlier, he boasted that he had achieved what Einstein could not. He read out loud at the meeting one of Einstein's private letters to him. In the correspondence, Einstein had called his earlier attempts at an affine theory "the tomb of his hopes."

"I think we can now exhume his hopes," Schrödinger said, "because I have lately been able to secure pretty strong observational verification of this part of the theory."[27]

Tagging the story with the headline "Einstein Had Failed," the *Irish Press* stated without substance that Schrödinger had succeeded where Einstein had "confessed failure." The article was misleading, as it implied that Einstein had long given up on unification, when exactly the opposite was true. He often conceded that earlier ideas were wrong while still clinging to the hope of eventual success.

What was the critical experimental evidence for Schrödinger's theory? There really wasn't any of substance. The alleged proof had to do with surveys of Earth's magnetic field. It is ironic that the compass, a device Einstein cherished as a child, became part of an attempt to render his ideas obsolete. And the surveys cited weren't even recent—one dated back to 1885 and the other to 1922. More up-to-date data was available and not even mentioned. For example, the very same month as Schrödinger's talk, geophysicist George Woollard published an article that systematically explored the magnetic and gravitational profile of North America.[28] Yet Schrödinger took his data from dusty old books.

Sometimes geophysicists find discrepancies between the expected and actual behavior of magnetic field lines. Those usually point to previously unknown magnetized structures beneath the surface, such as rock with greater than usual magnetite content. Therefore, if a geophysicist finds a compass dip anomaly, he or she might start to consider what kinds of underground formations might have caused it.

In general, readings of Earth's magnetic field fluctuate from time to time and from place to place. That's because the field is generated by a complex dynamo that is affected by the changing state of the magnetized materials within Earth's core, mantle, and crust.

Schrödinger, however, interpreted such discrepancies in a different way. He used them to purport that classical electromagnetic theory was slightly off in its predictions and (along with standard general relativity) must be replaced by his own unified theory. As the *Press* reported in a front-page story: "It is the reaction of the compass needle, recording variations in earth's magnetism intensity which has unexpectedly afforded proof of Prof. Schroedinger's great theory. It has done so in much the same way as the movement of the stars gave Einstein confirmation of the validity of the Theory of Relativity, which Prof. Schroedinger's new theory supplements and, in a measure, replaces."[29]

When Schrödinger developed his wave equation, he based it on known physical principles, such as the conservation of energy and the continuity of waves. Its success was due to matching well the precise spectral lines of atoms. When Einstein proposed his general theory of relativity, he based it on the principle of equivalence, a solid hypothesis grounded in measurements of how objects accelerate through space. It was tested via several independent means, including the bending of starlight by the Sun, which is hard to explain otherwise.

However, in the case of Schrödinger's general unitary theory being "confirmed" through the anomalous behavior of compass dips, there was neither strong theoretical justification nor significant experimental evidence. He had developed the theory through abstract mathematical reasoning, not on the basis of long-standing (or even hypothesized) physical principles. Furthermore, the evidence used to prove it found much simpler explanation in the natural variability of Earth's magnetic field. Even Schrödinger, at that point, viewed his theory as preliminary, not final. He would work on it for several more years before declaring victory once again. Yet the newspaper coverage had made it seem like the theory was a fait accompli, an undisputed major advance in science.

In August, Schrödinger wrote to Einstein with the electromagnetic "evidence" that his theory was right.[30] Einstein was skeptical. He replied in September listing other reasons Earth's magnetic field could be asymmetric, including the imbalance of ocean-covered regions in the Northern and Southern Hemispheres.[31] In October, Schrödinger wrote back, conceding, "You are probably right, as usual."[32]

Despite Einstein's critique, Schrödinger was undaunted. He excitedly explained to Einstein how he planned to extend the affine theory to include three fields: gravitation, electromagnetism, and the meson field (strong force). Gravitation and the meson field would be handled by the symmetric components of the affine connections, and electromagnetism would be relegated to the antisymmetric components. The idea sparked Einstein's curiosity and led to more discussions by mail.

Einstein continued to be pleased to have a kind of theory pen pal with whom he could toss around ideas. He wrote warmly to Schrödinger, "I greatly appreciate that you wish to inform me openly about your efforts. To some extent, I deserve it, because for decades I've been beating my bloody nose against a hard rock."[33]

With his newfound popularity (thanks to "What Is Life?" and his general unitary theory), warm correspondence with Einstein, and apparent strides in physics, Schrödinger was floating on air. Brilliant as he was, a kind of narcissism clouded his judgment. His desire for female admiration and the thrill he found in seduction made him ripe for more love affairs. He would have two within the next few years, each leading to pregnancies and baby girls.

The first was with a married woman named Sheila May Greene, an intellectual social activist and critic of the de Valera administration. They began their relationship in spring 1944. Sheila was pregnant by that autumn. On June 9, 1945, their daughter, Blathnaid Nicolette, was born. Sheila and her husband, David, would raise her—he on his own after the two of them separated. Aside from the daughter, one tangible result of the affair was a book of love poetry that Erwin wrote to Sheila and would eventually publish.

The second affair was with a woman named Kate Nolan (a pseudonym to maintain her privacy), a government worker who had become a friend of Hilde when they both volunteered for the Red Cross.[34] Their brief liaison resulted in a girl named Linda Mary Therese, born on June 3, 1946. Initially Kate, in a state of shock over the unplanned pregnancy, let the Schrödingers raise Linda. However, two years later, she decided that she wanted her daughter back. One day she saw Linda in a stroller, being wheeled around the neighborhood by the nanny the Schrödingers had hired. Kate lifted Linda out of the stroller and took her away. There was not much Erwin could do, since Kate was the legal mother. Kate brought her to Rhodesia (now Zimbabwe), where she would grow up. Linda's son (and Schrödinger's grandson) Terry Rudolph would be born there in 1973.[35] He became a quantum physicist and is currently working at Imperial College, London.

To Catch a Physicist

Living in neutral Ireland during the war, Schrödinger was spared the difficult moral decision of whether or not to contribute to military efforts. By remaining in Germany, Heisenberg, on the other hand, was in a position where it would have been hard to refuse. He had family ties to Heinrich Himmler, powerful chief of the Schutzstaffel or SS (the Nazi paramilitary corps) and the Gestapo (the secret police), which helped protect him from criticism that he had been too friendly with Jewish scientists such as Born. Those connections also helped him obtain leading scientific positions during the war. Rather than flinching or complaining, Heisenberg welcomed the opportunity to serve his country—even under a regime that he didn't support.

Much has been written about Heisenberg's wartime role guiding the Nazi nuclear program. After the war, he would downplay the team's efforts toward developing the bomb and play up its work toward peaceful aspects of nuclear energy. His physicist colleague Carl Friedrich von Weizsäcker would suggest that they dragged their feet and never wanted Hitler to get the bomb. They argued that, in a way, the German scientists behaved more ethically than the Allies because they didn't earnestly strive for nuclear weapons and never used them.

Heisenberg also accused Einstein of being hypocritical in transforming from a leading pacifist to a staunch supporter of the Allied war effort.

However, in 2002, unsent letters from Bohr addressed to Heisenberg were released in which he documented discussions they had held in Copenhagen in 1941 (meetings that became the basis of Michael Frayn's famous play *Copenhagen*). Bohr never sent the letters to Heisenberg because he didn't wanted to reopen old wounds. He recalled that Heisenberg had informed him that the Germans were actively working to build an atomic bomb and that they would ultimately prevail. Bohr had been shocked by Heisenberg's confidence. In September 1943, Bohr was forced to escape Denmark by fishing boat to Sweden, and then by military aircraft—arranged by Lindemann—to Britain, where he joined the Allied nuclear effort.

Monitoring Heisenberg and the German nuclear project became an important priority for Allied intelligence. Around the same time as Bohr's escape, Samuel Goudsmit (who with George Uhlenbeck had proposed the concept of quantum spin) was appointed to head the Alsos Mission, given the task of assessing Axis progress toward a bomb.

A most unlikely spy was Moe Berg, who was a mediocre major league baseball player and coach but a master of foreign languages and an expert in feigning scientific ability. One of his former teammates quipped that he "can speak in 12 languages but can't hit in any of them."[36] Berg joined the Office of Strategic Services, a predecessor of the CIA, in 1943, and soon was enlisted in a top-secret project to head off the Nazi nuclear effort.

After being briefed in the nuances of quantum and nuclear physics, Berg posed as a physicist and attended a Zurich conference in December 1944 where Heisenberg was scheduled to speak. Packing a pistol and a cyanide capsule, he had strict orders: if Heisenberg appeared to be making progress toward the bomb, Berg was supposed to assassinate him. If, on the other hand, he seemed to be doing innocuous research, Berg would leave him alone. Fortunately for Heisenberg, it was the latter. He spoke about scattering matrices in quantum physics, a subject that had little to do with bombs. Berg decided that it was safe to spare Heisenberg's life.

In 1945, as the Allies closed in on Berlin, Britain and the United States realized that any atomic secrets found by German scientists could fall into the hands of the Soviets. They launched Operation Epsilon to capture the top German nuclear physicists and bring them to England.

Heisenberg and nine others—including Hahn, von Weizsäcker, and von Laue—were taken to Farm Hall, a stately mansion near Cambridge, where they were detained for six months. Although isolated and under guard, they were made comfortable and treated well.

Riddled with hidden microphones, Farm Hall was a peculiar laboratory where the subjects were scientists themselves. The goal of its "experiment" was to see if relaxed, well-treated researchers, unaware that they were being monitored, would open up about what they had aspired to do and what they had actually found. When, in August, the Allies dropped atomic bombs on Hiroshima and Nagasaki in Japan, the scientists' reactions were recorded and carefully analyzed. By all measures, they were stunned that the Allied nuclear bomb project had proceeded so rapidly. While they had indeed been working toward a bomb, their efforts had been stymied by a lack of funding and Heisenberg's poor sense of experimental design. His thinking was too abstract for bomb making. Therefore, they had informed their superiors that building a bomb would take many more years of research and was not realistic in the near term.

After Heisenberg was released from Farm Hall, he resumed his academic career. With the war over and the Allies victorious, Germany returned to its prewar borders, with some adjustments. It was divided into four occupation zones, each administered by one of the Allies. Berlin was separately carved into four zones. Heisenberg settled in Göttingen, which had been assigned to the British sector.

Schrödinger was pleased to see Austria liberated and reestablished as a republic. However, it too had been divided into occupation zones, including a Soviet sector in the east. While he began to think about returning, because of the political situation he remained in Dublin for the time being—a waiting period that ended up being a decade. In the interim, he decided to step down as director of the School of Theoretical Physics, ceding the role to Heitler, while remaining as senior professor. His stated goal was to focus more on his research. However, it is reported that he was in a dispute with the maintenance staff at the institute and decided he had enough of administration.

A change also took place in their family life when Hilde and Ruth left for Austria. Anny had grown very close to Ruth during their years together on Kincora Road. She had become a kind of second mother to the child. When Hilde decided to return to Innsbruck and reunite with

Arthur, bringing Ruth with her, Anny was distraught. She sank into a deep depression, no doubt exacerbated by Erwin's continued serial liaisons with other women.

Einstein had absolutely no interest in returning to Germany. If he had, he would have found great devastation. Like much of central Berlin, his former apartment building in the Bavarian quarter had been destroyed. His lake house in Caputh, then part of the Soviet sector, had been appropriated. The face of most German cities had been fundamentally changed, requiring decades of rebuilding. However, most shocking had been the human death toll. The Nazis had systematically murdered millions of Europeans, including six million Jews. Millions more people had died in the war. Countless others were homeless, incapacitated, widowed, orphaned, or otherwise affected. Einstein would never forget or forgive such unspeakable horror.

Despite his great sadness and anger, Einstein resumed his efforts toward a unified field theory. Working with Ernst Straus, he began to explore what he called a "generalization of the relativistic theory of gravitation." Like Schrödinger's efforts in some ways, it involved tinkering with the affine connections that relate one point in spacetime to another and seeing how these changes affect the field equations. He had begun the project on his own, publishing his work as a single-author paper, but there was an error in his calculations, which Straus corrected. They published a joint paper in 1946.

By no means did Einstein consider his work on unified field theory complete. In his final decade of life, he would take an eclectic approach, treating various modifications of general relativity as options, rather than as the final word. Nonetheless, his monumental fame guaranteed an audience for virtually anything he would compose, no matter how abstract or preliminary. He would also have to grapple with contenders—particularly a certain old chum working in Dublin.

CHAPTER SEVEN

Physics by Public Relations

I don't like the family Stein;
There is Gert, there is Ep and there's Ein.
Gert's poems are bunk;
Ep's statues are junk,
And nobody understands Ein.

—AUTHOR UNKNOWN, in *Time*, February 10, 1947

After the war ended, Einstein's public image became decidedly more complex. Ironically, the bombing of Hiroshima and Nagasaki wedded him in the public mind to the war effort, despite the fact that he didn't even have clearance to do military research on atomic projects. The fact that mass turned to energy in the bomb blasts pinned the mushroom cloud image indelibly onto the theory of relativity. (Even today the idea that Einstein somehow was "father of the nuclear bomb" persists.) If Einstein was seen as brilliant before the war, its outcome seemed to bestow on him the powers of a superhero as well.

A reflection of this image was a strange rumor that appeared in Walter Winchell's popular newspaper column on May 23, 1948, entitled "Scientists See Steel Block Melted by Light Beam." He claimed to his millions of readers that Einstein was working with ten former Nazi scientists on developing an ultrapowerful death ray. As he reported, "The 11 scientists (led by Einstein) put on asbestos suits—and watched a beam of light. . . . A block of steel—20 by 20 inches—was melted as quickly as you turn on the switch in your home. . . . This new and secret weapon can be operated from planes and destroy entire cities."[1]

According to Einstein's FBI file, released in recent decades through Freedom of Information Act requests, the rumor was taken seriously enough to be refuted by the US Army's Intelligence Division, which stated that "this information could have no foundation in fact and . . . no machine could be devised that would be effective outside the range of a few feet."[2]

Einstein's FBI file also mentions worries that he would defect to the Soviet Union. Presumably part of the fear was that he would bring with him carefully guarded nuclear secrets. Such scrutiny certainly was overkill, given that he had never had clearance to work on the atomic bomb and knew nothing about its specifics.

Ironically, apparently unknown to the FBI, Einstein had a romantic relationship with a Russian woman, Margarita Konenkova, who was alleged to be a Soviet spy. Einstein met Konenkova in 1935, when her husband, acclaimed sculptor Sergei Konenkov, was creating a bust of Einstein for the Institute for Advanced Study. At some point they began an affair, which lasted until the end of the war. Historians learned about their relationship in 1998 when love letters from Einstein to Konenkova dating from 1945 and 1946 surfaced at an auction.[3] Anticipating the trend of name mashups, such as "Brangelina" for Brad Pitt and Angelina Jolie, Albert and Margarita nicknamed themselves "Almar." Around the time that the letters became known, a former Soviet spy asserted that Konenkova had been a kind of Mata Hari, trying to lure Einstein into spilling secrets about the US atomic program. She did indeed acquaint him with the Soviet vice consul in New York. However, to date no solid evidence has been furnished that she was a spy, let alone that she seduced Einstein into offering any clandestine military information—which he didn't have anyway. Luckily, the story was unknown to the tabloids of the day; Einstein was distrustful enough of the press already.

Einstein continued to find much of what was printed in newspapers about him ridiculous. Asked once by a Swiss paper about suitable reading for young people, his response showed a great disdain for the press: "A man who only reads the newspapers . . . reminds me of a very short-sighted man who is ashamed to wear glasses. He is completely dependent on the judgments and fashions of his age and he sees and hears nothing else."[4]

Nothing would prepare Einstein for the flood of attention when word came from Dublin that Schrödinger had seemingly beaten him to a unified field theory. To set the record straight, he would be forced to confront the inquiring journalists head-on. At least part of the impetus for the flurry of press reports was the situation faced by de Valera himself.

Under siege for the dire economic conditions facing Ireland, Dev hoped that DIAS, his Olympus of the mind, would shine. Schrödinger, his star player, needed to rack up more wins for Irish science. The *Irish Press,* his mouthpiece, felt obliged to root for the home team and trumpet its successes. Otherwise, the political opposition would be waiting in the wings to seize the moment. Watching closely for any missteps, they were hungry and eager to do so.

Dev's Falling Star

In the postwar era, after more than a decade in power, the *taoiseach*'s star had clearly fallen. Massive unemployment, food rationing, and emigration reminiscent of the time of the potato famine contributed to a growing sense that he was out of touch with common concerns. A new social democratic party, Clann na Poblachta, began to draw support away from Fianna Fáil. Debates in the Dáil grew nastier, with politicians attacking government policies needing to be reminded to refer to de Valera by his title.

De Valera remained heavily involved with DIAS. He and his party continued to cite its foundation as a major accomplishment, despite limited evidence that it had reached its goal of international prominence. For example, during the Irish general elections of 1948, his party listed the founding of the Dublin institute as one of its achievements.[5]

Amidst the social unrest, de Valera and the DIAS leadership decided to expand its mission by establishing a School of Cosmic Physics. Although adding new programs fell within the original mandate of the institute, he realized that he would need to request additional funding from the Dáil for the new school. In trying to crack open the treasury during a time of hardship, he was met with a blast of negative reactions. During the parliamentary debate, held on February 13, 1947, Fine Gael deputy James Dillon, a vocal critic of the de Valera administration, led the venomous attack.

"My submission is that it is being done for the purpose of securing cheap and fraudulent publicity for a discredited administration," argued Dillon. "I recollect a 'Life of de Valera' [in which] it was described that politics were an insufferable ordeal for him: that, really, his happiness was to be detached from all mundane affairs and left free to wander in the higher realms of mathematics, where few could follow him. This is the spectacle we are being entertained to now—himself and the cosmic physicist sitting over in Merrion Square rise in the cosmic ether while the ignoramuses of the Fine Gael Party, and the Clann na Talmhan Party and the Labour Party are occupying themselves with old age pensioners, and cows, and contemptible considerations of that kind."[6]

It is unclear if, in Dillon's remarks, "the cosmic physicist sitting over in Merrion Square" referred to Schrödinger or someone else. Regardless of who, besides de Valera, might have been the target of his barbs, clearly DIAS was under fire for being elitist. Schrödinger faced immense pressure to justify his salary and office during a time of desperate need.

Asymmetric Camaraderie

From their dialogue about wave mechanics in the early 1920s to their jaunts near Caputh in the late 1920s and discussions about quantum philosophy in the mid-1930s, Einstein and Schrödinger's friendship had deepened throughout the years. Their correspondence about unified field theory in the early 1940s had elucidated common interests that brought them even closer. However, arguably the time in which their theoretical goals and techniques were closest was the period from January 1946 to January 1947, when each sought to expand general relativity by removing its symmetry conditions. They searched for unity almost in unison. Only minor differences separated their ideas. During that year, they were collaborators in almost every sense, except that they didn't publish a paper together. Rather, their collaboration ended abruptly when Schrödinger, spurred in part by the pressure to perform for Dev, declared victory over Einstein in a dramatic announcement to the Royal Irish Academy.

Why would Schrödinger suddenly leave the duo and go off on his own? Although their theoretical interests were symmetric, Einstein's and Schrödinger's life situations at that time were markedly asymmetric. Einstein was little concerned about pleasing his superiors. By that point, the leadership of the Princeton IAS and the physics community in general treated him as a relic—more for show than for science—and he knew it. He also didn't have to worry about being productive to support his family, as Elsa was long gone. Rather, it was his own inner taskmaster that drove his unending struggle.

Schrödinger, on the other hand, felt rightly or wrongly that he still needed to prove himself, justify his salary, and maybe even earn a raise. The "What Is Life?" talk, mentioned in the international press, had boosted his standing. Doing something "Einstein-like" would help even more. Thus his mind had become focused on what Einstein was doing and whether he could either help out or do it better. He became something like a martial arts student carefully observing his coach's every move, trying to replicate each step and wistfully thinking about how he could one day come out on top.

Not that Schrödinger was in any way Machiavellian. He was genuinely interested in the project, which matched well his talent for mathematics. The last thing he wanted to do was to hurt or betray Einstein. Somehow he thought that he could impress Dev and the backers of the institute without Einstein personally being affected. He didn't realize until it was too late how his declaration of success would embarrass and offend his friend.

Through their correspondence, we can see how their ideas developed. On January 22, 1946, Einstein sent Schrödinger a letter describing how to generalize relativity by keeping the nonsymmetrical terms of the metric tensor. It was the work he had just completed with Straus. Recall that the metric tensor defines how distances in spacetime are measured, an extension of the Pythagorean theorem for curved spaces. It can be written as a 4-by-4 matrix with sixteen components. Normally, because of symmetry only ten of these are independent. However, Einstein decided to remove the symmetry, restoring the other six components as separate entities. His reason for adding more independent components to the metric tensor was to make room for electromagnetism—just as he had tried earlier with an extra dimension.

Einstein pointed out to Schrödinger that Pauli had raised objections to his new method. Pauli generally didn't like the idea of mixing symmetric and nonsymmetric components, believing that the system would not transform properly and would thereby be unphysical. Echoing a biblical verse, Pauli once told Weyl, "What God has put asunder, man should not join together."[7]

"Pauli stuck out his tongue at me," Einstein complained to Schrödinger.[8]

What was new about that? Pauli had been quick to criticize *every* technique that Einstein had tried. Unfortunately for Einstein, each time Pauli had been right. But could the "scourge of God" somehow be missing something in this case? Einstein wanted Schrödinger's guidance.

On February 19, Schrödinger wrote back with some suggestions. He showed how the metric tensor could be expressed in such a way that "Pauli would stop sticking out his tongue."[9] He also urged Einstein to incorporate the meson field (strong interaction), rendering the unification of the natural forces more complete.

The meson field turned out to be a sticking point between the two. Einstein didn't want to complicate matters by adding an extra interaction. He felt that it would be enough to find a mathematically reasonable theory of gravitation and electromagnetism that lacked pesky singularities. For Schrödinger, uniting two out of three of the then-posited interactions would be insufficient. He wanted a full trifecta that would conquer all known forces. Throughout the spring they argued about the issue, but each was stubborn and wouldn't bend.

On the other hand, Schrödinger at one point felt that Einstein was being too ambitious in trying to develop a singularity-free theory that would describe the complete behavior of electrons. As was his style, he turned to an animal metaphor to describe his feelings. "You are after big game, as they say in English," he wrote to Einstein on March 24. "You are on a lion hunt and I'm talking about rabbits."[10]

Gift from the Devil's Grandmother

Despite their differences in approach, Einstein and Schrödinger continued to grow closer. On April 7, Einstein started off a letter with a high compliment: "This correspondence gives me great pleasure, because you are my nearest brother and your brain runs so similar to mine."[11]

Schrödinger was thrilled and honored to be such a close confidant. One cannot imagine more glowing praise to a physicist than saying his brain works like Einstein's. Nothing would be more heavenly than reading those words in a letter from the great man himself. At another point Einstein called Schrödinger a "clever rascal," which made his head swell even further.

In their extensive dialogue, exchanging detailed letters once or twice a month, they had a number of laughs about the obstacles they faced.

One ongoing joke involved a comment Einstein made about a mathematical issue he confronted. He called it a "gift from the devil's grandmother." Einstein meant that in the sense of a hex—a creeping feeling that he was doomed to fail—but Schrödinger found the expression amusing. In his reply, Schrödinger furnished his own story. "It's been a long time since I've laughed out loud so heartily and uproariously as I did about the 'gift from the devil's grandmother,'" he wrote. "For in the preceding phrases you had accurately described the Calvary to which I had also gone, only to end up with something that was probably still as unsuitable as your result."[12]

Einstein replied, "Your last letter was indescribably interesting. And I was also quite stirred that you also paid such devotion to the 'devil's grandmother.'"[13]

Together they faced mathematical demons left and right. One issue that stymied the physicists was the concept of invariance. Standard general relativity has the ideal feature that simple kinds of transformations, such as a change in or rotation of the coordinate system, don't affect the physical results. However, some extensions of general relativity lack that invariance. Some of the components transform differently than others. This makes the theory less than ideal. It is like taking a smooth-driving car, adding a trailer, and hoping that if one turns to the right, the other follows at the same pace. Otherwise the whole thing might jackknife and fall apart.

By the end of 1946, the two were so close that Schrödinger tried to persuade Einstein to relocate to Ireland. It would be ideal for working together. Einstein politely declined, writing, "One does not place an old plant in a new pot."[14]

Sometime in January 1947, Schrödinger made what he saw was a major breakthrough. He found a simple Lagrangian that fit well with his general unitary theory to produce the field equations for gravitation, electromagnetism, and the meson field—or so he thought. Excitedly he prepared a report for the Royal Irish Academy, to be delivered at its January 27 meeting.

The Speech of a Lifetime

The Irish winter of 1947 was notoriously brutal. The bitter cold and snowfall made the severe fuel shortages sting even more. No wonder the government had become so unpopular. By the end of January, Dublin's temperatures had plunged into the freezing range and light snow had started to fall. The weather would get even worse as the winter progressed.

Despite a coating of snow on the ground, cyclists continued to plod through the streets of the city center. Schrödinger was undaunted by the weather, as he had a mission to complete. Biking along Dawson Street, a grand thoroughfare parallel to Dublin's main Grafton Street, he arrived at Academy House with the "key to the universe" in his sack: a simple combination of symbols that could be scrawled on a postage stamp but which he believed was the Lagrangian representing everything in the universe. Plug that Lagrangian into the equations of motion developed by Hamilton and all the forces would miraculously appear.

Hamilton's spirit haunted the stately brick building. In 1852, the year the academy moved to 19 Dawson Street, he was Ireland's leading scientific thinker and a steady presence at meetings. Greatly interested in the relationship between time and space, he would have been fascinated to see how physicists connected them in mathematical theories. As he once remarked, "And how the One of Time, of Space the Three, Might in the Chain of Symbols girdled be."[15]

The academy's meeting room, designed by noted architect Frederick Clarendon, was the epitome of elegance. Large chandeliers hung from the high ceilings, supplementing the hazy illumination from tall windows situated above the balconies. Reminding members of the value of past scholarship, bookcases packed with weighty tomes lined the walls. Each series of lectures, carefully recorded for posterity in the academy's proceedings, added another work to the collection.

The *taoiseach* took his place in the hall, along with about twenty other attendees, including students and professors. No doubt he was happy to be there, rather than debating his angry opposition in the Dáil. His presence virtually guaranteed press coverage. Reporters from the *Irish Press* and *Irish Times,* tipped off that the meeting might be newsworthy, sat wide-eyed and eager for a story.

The academy's president, Thomas Percy Claude Kirkpatrick—a physician, bibliophile, and historian of medicine—stepped up to the podium. He too had arrived by bicycle, as he didn't own a car. Kirkpatrick introduced a new member, the Earl of Rosse, and the first speaker, botanist David Webb, who lectured about a type of plant native to Ireland.

Meeting Room of the Royal Irish Academy, where Schrödinger delivered many of his key lectures. *Photographer and date unknown; by permission of the Royal Irish Academy* © *RIA*

Then it was Schrödinger's turn to speak. The hall grew silent and all eyes were on the Austrian Nobelist.

"The nearer one approaches truth, the simpler things become," Schrödinger began. "I have the honor of laying before you today the keystone of the Affine Field Theory and thereby the solution of a 30 year old problem: the competent generalization of Einstein's great theory of 1916."[16]

Reporters carefully scribbled notes about the new scientific revolution. Visions of headlines danced in their heads. They hoped they could somehow get the math straight enough to convey its importance to readers.

Schrödinger explained how Einstein and Eddington had almost stumbled on the correct Lagrangian, the square root of the negative of the determinant of the Ricci tensor, but he was the one who had actually made it work. (Recall that the Ricci tensor is a way of describing the curvature of spacetime; its determinant is a way of summing its components.) Schrödinger pointed out that the key difference between earlier efforts and his was that he used a nonsymmetrical affine connection. Unnamed colleagues had tried to dissuade him, but he had stuck to his guns.

Schrödinger made use of an animal analogy (his favorite kind) to justify his use of an affine connection that wasn't symmetric and thereby included extra, independent components. "A man wants to make a steed take a hurdle," he said. "He looks at it and says: 'Poor thing, it has four legs, it'll be very difficult for him to control all four of them. I know what I'll do. I'll teach him in successive steps. I'll bind his hind legs together. He'll learn to jump with his fore legs alone. That'll be much simpler. And then we'll see, later on perhaps he'll learn with all four.' This describes the situation perfectly. The poor [affine connection] got his hind legs bound together by the symmetry condition, taking away 24 of its 64 degrees of freedom. The effect was, it could not jump; it was put away as good for nothing."[17]

At the end of the talk, Schrödinger made the stunningly ambitious prediction that his theory would explain why a rotating mass, such as Earth, has a magnetic field. Since 1943 his goals had ballooned from elucidating anomalies in Earth's magnetic field to explaining the whole thing. Talk about overreaching! He knew little about geomagnetism and seemed not to be aware of the strides made in understanding it through a model of Earth's core.

For example, in 1936 Danish geophysicist Inge Lehmann demonstrated through an analysis of seismic waves that Earth has both an inner core and an outer core. In 1940, American geophysicist Francis Birch developed a model of Earth's magnetism based on suppositions about the high-pressure behavior of iron in its interior. While his model was rudimentary and inaccurate, it represented a reasonable starting point for illuminating the source of Earth's magnetism. Given that history, Schrödinger's shot at explaining geomagnetism not only missed its target but wasn't even aimed at the right firing range.

A Dragon in Winter

At the end of the meeting, Schrödinger rushed out on his bicycle, evading curious reporters. He pedaled hard through the snow, weaving through traffic as quickly as he could. The journalists caught up with him at his home on Kincora Road. He handed them copies of his talk and sent them additional pages of explanation in language for laypeople. The news stories were undoubtedly on their way—national reports, maybe even international accounts.

Schrödinger's press release, "The New Field Theory," began with a historical account of ideas about particles and forces, starting with the ancient Greeks and ending with Einstein. He showed how a constant thread has been the desire to describe force and matter through geometry. That was the illustrious background to his own efforts. His chronicle seemed to imply that he would be the logical successor to the Greeks and Einstein. After describing the essence of his theory, he reiterated how Einstein and Eddington would have come up with the same thing in the 1920s had they been more open-minded. He mentioned his belief that he was almost certainly correct. The proof would be tests of Earth's magnetic field, which he believed could be explained only through his theory.

The *Irish Press* reported the next day that Schrödinger's address was history-making. "The theory should express everything in field physics," it quoted him as saying. Following a summary of his lecture, the report included a personal interview with him in which he was asked to explain his theory in simpler language. He replied, "It is practically impossible to reduce the theory that the man in the street can understand. It opens up a new field in the realm of Field Physics. It is the type of thing we scientists should be doing instead of creating atomic bombs. This is a generalization. Now the Einstein theory becomes simply a special case. Just as a stone thrown directly upwards is a special case in the general idea of a parabola."[18]

Asked about Einstein's earlier rejection of a version of this theory, Schrödinger replied that it was a good lesson to younger physicists that even the most brilliant scientists could be wrong. In other words, he was claiming that he had been savvy enough to ignore Einstein's authority and proceed on his own toward the correct solution. Conceding that, alternatively, he could be the misguided one, Schrödinger said that in that case he'd seem like an "awful fool."

The international media soon picked up the *Irish Press* story. For example, on January 31, the *Christian Science Monitor* reported Schrödinger's assertion that he had beaten Einstein to a unified field theory and thereby fulfilled a thirty-year quest.[19]

After his initial burst of confidence, Schrödinger soon began to have nagging doubts about how his bravado would be perceived. What would Einstein think when he found out? Surely he would understand the circumstances—the Dublin Institute for Advanced Studies being under siege and desperately in need of funding; the presence in the audience of de Valera, whom he had to impress; the hounding by reporters. It was just a scholarly talk, after all. He had made his claims within the academic context; it was the press that set out to disseminate them. Those were some of Schrödinger's justifications for his actions.

On February 3, he composed a letter to Einstein, explaining his new results and alerting him to the press situation. Warning Einstein that reporters inquiring about his reaction to the new theory might soon hound him, Schrödinger offered a weak apology. The salary and pension situation at the Dublin IAS, Schrödinger explained, was so bad that he had needed to boast a bit to bring more attention to the institute. In other words, he had somewhat exaggerated the importance of his findings for the sake of needed publicity for his cash-starved institute.

At the end of the letter, Schrödinger ruminated about what he would do if his determinant-based Lagrangian was wrong. "I'm going to sleep with the determinant and wake up with it," he wrote. "There is simply nothing else sensible. . . . If that is not right, then I'll let myself be Iguanodon, say 'Cold, cold, cold,' and stick my head in the snow."[20]

Schrödinger explained to Einstein that an iguanodon is a character from a story by Kurd Lasswitz. While he didn't elaborate on that literary reference, let's see what he likely meant. Lasswitz was a noted science fiction writer. In his novel *Homchen—Ein Tiermärchen aus der oberen Kreide* (Homchen: an animal tale from the Upper Cretaceous), the iguanodon is a long-necked dragon from prehistoric times. He lives among thick ferns, accustomed to the heat of the Sun. One day, he is disturbed to find that the weather outside is freezing cold. He sticks his neck out of his lair but quickly draws it back, muttering, "Cold, cold, cold." Because of the climate change, he is stuck without breakfast until it gets warmer. Who knows how long he will need to wait?

Indeed, in the snowy winter of 1947, Schrödinger was like a dragon that roared stridently but then had to retreat. The flames of his grandiose assertions burned his relationship with one of his closest friends. His collaborative efforts investigating unified field theories went up in smoke. Einstein stopped responding to his letters for the time being. Exactly as he feared, Schrödinger would be left out in the "cold, cold, cold."

Dissing Dublin

One of the international news stories bounced back to Dublin and hit its pride hard. Under de Valera, Dublin was trying to position itself as a center of scientific research. Yet a story published on February 10 in *Time* not only ignored those efforts but seemed to paint the city as the antithesis of science. The article began: "Last week, from nonscientific Dublin, of all places, came news of a man who not only understands Einstein, but has bounded like a bandersnatch far ahead (he says) into the hazy, electromagnetic infinite. . . . If so, he has scored a scientific grand slam."[21] The piece included Schrödinger's proposed Lagrangian and related formulas at the top of the page, mentioning that "to the non-scientific, it looks like incomprehensible doodling."

The reporter's curt dismissal of Irish science caught the eye of Dublin-born mathematician John Lighton Synge, who was then a professor at the Carnegie Institute of Technology in Pittsburgh. Synge penned a letter to the editor, published in the March 3 issue, calling him out for allowing the reference and pointing out that Hamilton was from Dublin.[22]

Rather than acknowledging Synge's example of a well-known Dublin scientist, the editor ventured into the personal. In a rebuttal following the letter, he raised the case of Synge's uncle, playwright John Millington Synge, as the reason Dublin should be associated with authors and not researchers. "Let Dublin-born Higher Mathematician Synge call to mind his city's great ghosts (among them, his uncle's—author of *The Playboy of the Western World*), and admit that Dublin is a writers' town."

Synge undoubtedly wanted to distinguish himself from his uncle and clarify that Dublin is the home of talented people of many different professions. The editor's reply shows how hard it is to throw off stereotypes.

Interestingly, the following year, Synge would be appointed to the Dublin Institute for Advanced Studies, where he would work for many years alongside Schrödinger. There he would make significant contributions to the study of general relativity, so much so that his biographer would call him "arguably the greatest Irish mathematician and theoretical physicist since Hamilton."[23]

Irish newspapers took note of the debate over Dublin's scientific merits. The *Irish Times* praised Synge as "a brilliant mathematician."[24]

Another Irish publication, the *Tuam Herald*, made reference to the parliamentary ruckus about the School of Cosmic Physics. After recapping the *Time* story and Synge's comments, it concluded, "The attitude of some of our Deputies in the recent Dáil debate on Cosmic Physics provides plenty of food for thought."[25]

Indeed, the discussions on both sides of the Atlantic about whether Dublin was "non-scientific" demonstrates the power of the fallacy de Valera wanted to dispel by establishing DIAS and recruiting Schrödinger. Despite his efforts, it seems that he fell short of his goal of an Irish scientific renaissance acclaimed worldwide.

Einstein's Retort

Naturally, the public would be interested in the sage of relativity's own views on whether he had been beaten to his goal of unification. Reporter William Laurence, the *New York Times* correspondent who generally dealt with Einstein in his later years, sent him a copy of Schrödinger's paper and press release to gauge his reaction. Laurence also sent copies to Eugene Wigner, Robert Oppenheimer, and other prominent physicists. In his note to Einstein, Laurence said, "If, on perusing the papers, you find yourself in agreement with Dr. Schroedinger, I would deeply appreciate a statement from you to that effect."[26]

The *Times* published three articles about the alleged breakthrough, including a remark by Einstein that he "declines comment" (for the time being, as it turned out).[27] Another piece, describing the talk, included in its headline "Einstein's Theory Reportedly Widened: Scientist in Dublin Claims He Has Achieved Unified Field Theory Sought for 30 Years."[28] The third piece mentioned that while Schrödinger could be right, he was "aware of the pitfalls in his path."[29]

Soon thereafter, another press group, the Overseas News Agency, independently sent Einstein another copy of Schrödinger's article. Rubbing salt in the wound, the agency's managing director, Jacob Landau, similarly asked for his opinion about "the merits of the formula and its implications."[30]

Judging by his reaction, Einstein was furious. With the help of Straus, he composed his own statement to the press. While the tone of the statement starts off as neutral and scientific, by the end it is caustic. Einstein wrote: "The foundations of theoretical physics are not determined at present. We are striving to find first a usable (logically simple) basis of theoretical physics. The layman is naturally inclined to consider the course of development as such that the basis is obtained from the facts of experience by a gradual generalization (abstraction). This however is not the case."

After explaining how Schrödinger's theory is simply a mathematical exercise (and not particularly a good one) rather than a real physical result, Einstein concludes by scolding the press: "Such communiqués given in sensational terms give the lay public misleading ideas about the character of research. The reader gets the impression that every five minutes there is a revolution in science, somewhat like the coup d'état in some of the smaller unstable republics. In reality one has in theoretical science a process of development to which the best brains of successive generations add by untiring labor, and so slowly leads to a deeper conception of the laws of nature. Honest reporting should do justice to this character of scientific work."[31]

Einstein's comment about press coverage was apt. However, it also applied to the reporting of his own unified field theory attempts, which in many cases were treated as breakthroughs rather than just works in progress. For example, during the media hullabaloo over his 1929 theory of distant parallelism, instead of quashing speculations he added to them with his own public statements about its importance.

After Einstein's critical response was published in venues such as *Pathfinder*, a Washington-based news journal, as well as in the *Irish Press*, Schrödinger issued his own press statement, framing the matter as an issue of academic freedom: "Surely Professor Einstein is the last to dispute an academician's right of reporting to his Academy and giving his opinion freely."[32]

As Anny recalled, there was even talk of lawsuits, with each thinking of charging the other with plagiarism. When Pauli found out, he decided to mediate. He cautioned them about the bad publicity such a legal action would stir up. "Besides," he noted, "I really don't see what the whole fuss is about. This theory is ill conceived. If you connected *my* name with it in any fashion then I would have a right to sue *you*."[33]

Schrödinger soon decided that it would be unwise to further the dispute. He was in enough trouble with his friend. Things had gotten out of hand. He started to call the incident the "Einstein *schweinerei* [mess]."

Even though Schrödinger refrained from continuing the argument, a certain humor writer decided to go to bat for him. Under his pseudonym, Brian O'Nolan wrote a scathing column accusing Einstein of snobbery. "Do you know I don't like that speech at all," he commented. "First of all, note the assumption of the druidic mantle by the opening sneer. . . . I am, forsooth, a *layman*. And the layman is *naturally* inclined to consider something quite stupid [such as] gold grows on trees. . . . It is just abuse, nothing more."[34]

Like Schrödinger, Einstein also set the argument aside. (He didn't respond to the O'Nolan piece, which he probably didn't even hear about.) However, it would be three more years until he resumed his correspondence with his old friend.

Milestones

In 1948, Princeton physicist John Wheeler, who lived near Einstein and often visited him, brought him exciting news. Wheeler's brilliant student Richard Feynman had developed a unique approach to quantum mechanics, called "sum over histories," that generalized Hamilton's least-action principle to the study of how photons transfer between electrons and other charged particles to generate the electromagnetic force. In creating a force, the photon acts as what is called an "exchange particle." (Its existence is required through Weyl's gauge theory of electromagnetism.) Unlike in classical mechanics, in which particles travel along unique paths, Feynman showed how in quantum interactions all possible paths are taken, weighted by their probabilities to create a net result.

We can understand the difference between classical mechanics and Feynman's sum over histories through an analogy involving a boy wearing boots walking home from school. Suppose he has the option of three different possible routes: a quick way through sand, a slightly longer way through mud, and an even longer way through gravel. In classical mechanics, he would choose the most efficient route and his boots would be covered in sand. Contrast that with the quantum version, in which the result would be a sum over histories. In that case, his boots would have a lot of sand on them, but also a measure of mud and a bit of gravel. It would be like he took all three routes at once, but somehow "most of him" took the quickest path.

One initial issue with Feynman's method was the appearance of unwanted infinite terms. However, he—and, independently, physicists Julian Schwinger and Sin-Itiro Tomonaga—developed a way of canceling out these infinities, called "renormalization." Renormalization involves arranging terms such that the additions and subtractions leave a finite sum.

Feynman, Schwinger, and Tomonaga's contributions, called quantum electrodynamics or QED, opened the door to a greater understanding of particle interactions. Though designed for the electromagnetism interaction, their methods would eventually be modified to characterize the weak and strong nuclear forces. It would prove to be a decisive step toward a standard model of the forces—a way of understanding electromagnetism, the weak and the strong interactions, through a unified explanation.

Einstein had little interest in such ideas. As Wheeler recalled, he was not impressed by Feynman's sum over histories notion. The issue was its reliance on probability. "I can't believe that God plays dice," Einstein told Wheeler, "but maybe I earned the right to make my mistakes."[35]

That year, Schrödinger (along with Anny) became an Irish citizen. He was by all measures content in his adopted country, save his longing for the Austrian mountains—and, of course, for Hilde and Ruth. The only hitch was that his mentor was no longer the *taoiseach*. Dev was forced to step down after the February 1948 elections, when the opposition parties assembled a parliamentary coalition that drove Fianna Fáil out of power. The changing of the guard proved that Irish democracy was healthy. Soon Ireland would officially become a republic—as it had been in essence since the late 1930s.

Just as 1949 began, several months before his seventieth birthday, Einstein needed abdominal surgery. He checked into Brooklyn Jewish Hospital on New Year's Eve, was operated on, and rested up for a number of days. As he left the hospital via its rear exit, hordes of paparazzi swarmed around him and asked him to pose. He vehemently refused, screaming, "No! No! No!"[36] But the photographers still wouldn't leave him alone, and eventually he had to call the police to be escorted away.

A highlight of his birthday celebration that year was a visit by a number of children made homeless by the war, including a young distant relative, Elizabeth Kerzek, age eleven. William Rosenwald, chair of the United Jewish Appeal, brought the refugees to Einstein's house. Rosenwald pledged to Einstein that he would try to find new homes for all the refugees by the end of the year. As a refugee himself, Einstein was a strong advocate for finding new homes and positions for displaced Europeans, and wrote countless letters in support of such efforts.

The Institute for Advanced Study, directed at that time by Oppenheimer, who had assumed the post at the end of the war, honored its most famous scientist with a well-attended conference, cosponsored by Princeton University. Weyl, who had long abandoned the pursuit of unification for pure mathematics, was one of many thinkers paying tribute. Physicists had begun applying Weyl's gauge theory to the world of particles with excellent results.

Bohr could not attend the conference but sent recorded congratulations. He had come to relish his dialogue with Einstein, which he saw as a kind of "trial by fire" for the tricky aspects of quantum physics. He felt that his and others' responses to Einstein's hard-hitting queries helped render quantum philosophy stronger.

Another speaker, physicist I. I. Rabi, predicted that the effect of gravitation on atomic clocks would be accurately tested by the time of Einstein's eightieth birthday. He was not so far off. Although Einstein would not live to that age, in 1959, when he would have turned eighty, Harvard physicist Robert Pound, along with his student Glen Rebka, conducted experiments that successfully measured the gravitational redshift—the general relativistic effect of gravity on light's frequency. It would be another triumph for Einstein's theory.

Wigner, another attendee, also lauded Einstein. Later in life he would develop a strong interest in the EPR and Schrödinger's cat thought experiments and examine the conundrums of quantum measurement theory. His "Wigner's friend" paradox would extend the cat dilemma by imagining a friend who opens the box and observes the cat but doesn't report the results. In this case, from the perspective of an outside observer, Wigner wonders if the friend would be in a mixed quantum state of shocked and relieved until he conveyed the outcome. The thought experiment would further highlight the role of consciousness in the orthodox interpretation of quantum mechanics.

Newspapers also honored Einstein, pointing out his ongoing search for cosmic truth. The *New York Times* aptly noted that he "will continue to seek until the end of his days for . . . an all-embracing concept that would also include gravitation and electromagnetism amid the vast forces within the nuclei of atoms that hold the universe together in one fundamental law."[37]

On that account, by the end of the year, he was already sensing success once again. The septuagenarian still had the spark to ignite the world's imagination. Pushing contenders aside, it would be Einstein's turn in the spotlight once more.

CHAPTER EIGHT

The Last Waltz:
Einstein's and Schrödinger's Final Years

> Of what is significant in one's own existence one is
> hardly aware, and it certainly should not bother the
> other fellow. . . . The bitter and the sweet come from the
> outside, the hard from within, from one's own efforts.
> For the most part I do the thing which nature drives me
> to do. It is embarrassing to earn so much respect and
> love for it. . . . I live in that solitude which is painful in
> youth, but delicious in the years of maturity.
>
> —ALBERT EINSTEIN, "Self-Portrait"

Einstein spent the months leading up to his seventy-first birthday much the same way as he had before his fiftieth: unveiling and promoting a new unification theory. To coincide with that occasion, Princeton University Press decided to release in March 1950 a revised edition of *The Meaning of Relativity*, a book based on talks he had given at Princeton University in May 1921 about the subject. The updated version would include an appendix in which Einstein explained his "generalized theory of gravitation" in popular form.

The last thing Einstein needed on that occasion was another distracting fight in the media. He did not have to worry about Schrödinger, who was humbled and on his best behavior. Vexed, no doubt, by the silence between them, Schrödinger had come to realize how foolish it had been to jeopardize a friendship for a fleeting chance at glory. However, Einstein could not escape controversy. A row occurred behind the scenes about premature mentions of the new material.

Datus Smith and Herbert Bailey, respectively director and editor of Princeton University Press, had everything timed perfectly for the unveiling of Einstein's latest theory. They planned to issue a news release in February, when copies of the book would be available. The public would learn about the purportedly groundbreaking new vision of nature by buying the book and perusing its appendix.

However, around Christmas 1949, Smith and Bailey found out that Einstein had also arranged independently with *Scientific American* to publish a piece that he had written about the generalized theory. *Scientific American* was planning to announce that shortly. The last thing the press wanted was for people to ignore the book in favor of the article. Consequently, they decided to bump up their announcement.

Shortly after their press conference, they were astonished to read a piece in the January 9 issue of *Life* written by Lincoln Barnett, who had recently authored a book called *The Universe and Dr. Einstein*.[1] Not only did the article explain Einstein's generalized theory in layman's language, scooping the planned appendix, but it failed to mention the new edition or Princeton University Press. Instead, it suggested that Einstein's theory had already been published. That was partly true; he had published other versions, but he had modified it in the interim. Smith and Bailey were worried that readers would be confused and that the momentum for the book would taper off.

After Smith dashed off an angry letter to Barnett about the lack of acknowledgment, Barnett wrote back apologetically, explaining that he had inadvertently omitted mention of the new book.[2] *Life* had wanted to beat *Scientific American* to what it saw as a hot story. Moreover, he had gotten the information about Einstein's new theory independently, through an American Association for the Advancement of Science conference in which an earlier version had been on display. Finally, he had thought that there would be other mentions of the theory in *Life* and that those would cite the book. Smith accepted his sound explanation and gracious apology.[3]

To complicate Smith and Bailey's lives even further, around that time Einstein phoned and told them the equations for his generalized theory could be expressed in simpler form. He insisted that they stop production of the book until he could revise the appendix, which needed to be translated from German into English by Bargmann's wife, Sonja. The press complied, even though it undoubtedly cost them some money. It was Einstein, so what else could they do? After the book was printed, Einstein found some errors in his calculations, which were corrected in an errata slip inserted into each copy.

There was yet another wrinkle to the story. In mid-January Einstein received a letter from a certain Frances Hagemann of Maplewood, New Jersey. She alleged that a phrase used in the *Life* article, "single harmonious edifice of cosmic laws," was her copyrighted property and that he had stolen it via the Commission on Atomic Energy.

"This is to warn you to keep your hands off my property," she wrote. "I haven't read your book yet but when I do if I find any infringement upon my Copyright I shall prosecute you to the fullest extent of the Copyright Law."[4]

Hagemann also sent a copy of the letter to Bailey. Bailey wrote back and explained that the phrase in question was *Life*'s, not Einstein's.[5] Hagemann still wasn't satisfied. She huffily responded to Bailey, with a copy to Einstein, that it was her ideas that were copyrighted, not just her words.[6] Records don't indicate whether she ever formally launched a legal complaint.

Word about the theory also reached the international press. In the *Irish Times,* a reporter decried that most people weren't educated enough to understand Einstein's new theory, except for a few savants such as Schrödinger. As the journalist wrote: "Unfortunately, Dr. Einstein is in a field by himself, and only a handful of men in other parts of the world can succeed even in scrambling through the hedges with which it is surrounded. . . . Ireland is fortunate in as far as one of her citizens, Dr. Schroedinger, belongs to the select band of human beings who may be able to understand and, what is more, to explain some aspects of the new theory."[7]

The *New York Times* lauded the new work as Einstein's "master theory." "His latest intellectual synthesis," it speculated, "may reveal to man vast forces beyond imaginings still hidden from view."[8]

It is remarkable that in Einstein's seventy-first year, more than a quarter century after his last groundbreaking publications, his mere proposal of a set of equations for unification that had met no experimental test created such a stir. Each Einstein theory, credible or not, was like sweet nectar to swarms of reporters and physicist wannabes, who rushed in to get—and sometimes even had to battle for—a taste.

From the standpoint of the mainstream physics community, in contrast, Einstein's successive unification attempts seemed increasingly ludicrous in the face of what they omitted about the known world of particles. A host of novel subatomic constituents, such as muons, pions, and kaons, had turned up in cosmic ray data, and Einstein's theories didn't even address them. He consistently ignored nuclear forces.

Robert Oppenheimer, for example, though very fond of Einstein and a great admirer of his early, seminal work, found his latter-day efforts absurd and unbecoming. As Oppenheimer wrote: "I think that it was clear then . . . that the things that this theory worked with were too meager, left out too much that was known to physicists but had not been known much in Einstein's student days. Thus it looked like a hopelessly limited and historically rather accidentally conditioned approach. Although Einstein commanded the affection, or, more rightly, the love of everyone for his determination to see through his program, he lost most contact with the profession of physics, because there were things that had been learned which came too late in life for him to concern himself with them."[9]

Humbled and Hopeful

Schrödinger felt awful about the row between him and Einstein three years earlier. To help make amends, he was generous in praising Einstein's unification efforts, while dismissing his own work.

"I was among those who have made such attempts without attaining something really satisfactory," conceded Schrödinger. "If he has now succeeded in doing so, it is certainly very important."[10]

Despite Schrödinger's desire for reconciliation with Einstein, there continued to be important differences between their standards for what constitutes a complete theory. Unlike Einstein, Schrödinger continued to press for the inclusion of nuclear forces. While Einstein seemed to have given up making experimental predictions, Schrödinger always emphasized their importance, even if his sense of what constituted evidence could be way off base. He kept going back to the example of Earth's magnetic field, even if he didn't really understand geophysics. Also, as the originator of the wave equation, Schrödinger was more likely than Einstein to emphasize the predictive success of standard quantum mechanics. Finally, dating back to his earliest papers on general relativity, published in 1917, Schrödinger maintained an active interest in the cosmological constant term, which Einstein had discarded.

Einstein had set aside the cosmological term in light of Hubble's discovery of cosmological expansion. Schrödinger, on the other hand, thought the term was essential, albeit small. He made a case for the cosmological constant in his 1950 book *Space-Time Structure,* a comprehensive survey of general relativity and related theories. He argued that one advantage of his affine theory was that it explained the source of the cosmological constant in a natural way and mandated that it have a value that was small but not zero.[11] Schrödinger's advocacy of a small but nonzero cosmological constant was certainly prescient. It matches well today's picture of an accelerating universe, propelled by unknown dark energy. Somehow his hunch turned out to be right on the mark.

In his book, Schrödinger also addressed the possibility that no solutions would be found for unified theories, but did not see that as an impediment. He also noted that if classical solutions were found, they might not match the quantum properties of the particles in question.[12]

Unlike Einstein, Schrödinger believed that generalizations of general relativity alone would not be enough to produce realistic particle solutions. He acknowledged that simple wavefunctions, solutions of his own wave equation, were more useful in revealing the nuances of quantum mechanics.

Take It to the Supreme Court

By autumn 1950, Einstein and Schrödinger's correspondence had resumed. Perhaps they realized how much they valued each other as sounding boards. Schrödinger continued to be extremely careful not to offend his dear friend. He had learned not to mouth off about the superiority of his theories.

Einstein was continuing to tinker with his generalized theory. In a letter to Schrödinger dated September 3, he acknowledged that his efforts might seem a bit quixotic. "All this has the whiff of good old Don Quixote," he wrote, referring to one of his mathematical suppositions, "but if you want to maintain the requirement of representing reality, there's no other choice."[13]

Their discussions turned to the unsatisfactory aspects of quantum measurement—a favorite mutual topic. Schrödinger's ever-shifting interests had snapped back to philosophy. He was eager to show that in the context of history, the orthodox interpretation of quantum mechanics would someday become a relic. He set down his views in a 1952 paper, "Are There Quantum Jumps?," which compared quantum discontinuity to the discarded Ptolemaic astronomy of epicycles that had been replaced by the Copernican system. He sent Einstein a copy of his paper, no doubt hoping for an enthusiastic reaction.

Shortly thereafter, unified field theories based on the affine concept started to come under attack. Several papers were published in 1953, including articles by physicists C. Peter Johnson Jr. and Joseph Callaway, demonstrating that Einstein's generalized theory—and, by extension, Schrödinger's work—would not yield the proper behavior of charged particles in nature. Einstein was quick to rebut the critiques, but Schrödinger became further disillusioned.

In May 1953, after receiving a copy of Einstein's latest ideas, Schrödinger offered a bit of constructive criticism with a few mathematical suggestions. Hoping not to upset Einstein, he started off the letter by writing, "Please do not be angry about my recalcitrance."[14]

Einstein's reply, in June, offered humor about their banter. "We've argued a lot, and unsuccessfully, about the naturalness of the affine theory. Only dear God can sit in judgment about intuitive decisions. As is the case here for the Supreme Court, He doesn't have to deal with such appeals."[15]

Bohm's Spin on Quantum Measurement

Many physicists who spent time in Princeton during the 1940s or early 1950s have their own personal Einstein stories. Some saw him walking through the town, perhaps with his assistants by his side. Others attended one of his lectures, typically in German. The lucky few who had the chance to meet him and have a personal discussion bear indelible memories of those precious moments—stories they have undoubtedly told their friends and family members many times.

Amherst College physicist Robert Romer has written about his "half hour with Einstein"—an invited visit to Einstein's house in February 1954. The meeting was pleasant and memorable. "Ms. Dukas welcomed me and showed me upstairs to Einstein's small and cluttered study," he recalled. "And there was Einstein, looking 'just like Einstein': khaki trousers, gray sweatshirt, dressed about as fashionably as I dress now."[16]

One of Romer's poignant memories was a discussion they had about the EPR thought experiment. He recalled Einstein asking, "Do you really believe that if someone here measured the spin of an atom, it could affect the simultaneous measurement of the spin of another atom way over there?" while pointing down Mercer Street. In hindsight, Romer was surprised that Einstein expressed the experiment in terms of spin, rather than position and momentum, as in the original paper. It seemed to be an early reference to the spin version of EPR, as introduced by physicist David Bohm. Bohm would publish that variation in a 1957 paper with Yakir Aharonov.

Einstein came to know Bohm when he was an assistant professor at Princeton in the late 1940s. Bohm had a strong interest in quantum mechanics and decided to write a textbook about the subject. After publishing the book, he began to doubt aspects of its orthodox explanation, including "spooky action at a distance." He conveyed his doubts to Einstein, and they had many fruitful discussions about the logical gaps in quantum theory. He decided to develop an alternative deterministic explanation using hidden variables: undetected, behind-the-scenes factors. By then, he had been forced to leave Princeton because of his refusal to testify before the House Un-American Activities Committee during the McCarthy-era witch-hunt for suspected communists. With Einstein's help, he obtained a new position at the University of São Paulo in Brazil. There he continued his explorations of a causal replacement for standard quantum mechanics. The result was a theory that harked back to the ideas of de Broglie and Schrödinger in the 1920s that the wavefunction was physically real, not just a repository for probabilistic information about particles. In 1927, de Broglie had published a deterministic interpretation of quantum mechanics based on realistic waves that guide particle behavior, calling them "pilot waves." Therefore, sometimes the ideas of de Broglie and Bohm, although developed independently, are lumped together as the "de Broglie-Bohm theory."

The Bohm-Aharonov version of the EPR thought experiment imagines two electrons from the same energy level propelled in different directions. Pauli's exclusion principle guarantees that the electrons must have opposite spin states: if one is spin up, the other is spin down. Until a measurement is taken, it is impossible to know which is which. Therefore, the two electrons form an entangled quantum state that is an equal mixture of both possibilities: up-down and down-up. Now suppose an experimenter measures the spin of one of the electrons using a magnetic apparatus and another researcher immediately records the spin of the other. According to the orthodox quantum interpretation, the system would instantly collapse into one of its spin eigenstates, either up-down or down-up. So if the first electron's reading was spin up, the other would automatically be spin down. In the absence of an interaction through space between the two, how would the second electron instantly "know" what to be?

In 1964, physicist John Bell would explore this issue further by developing a mathematical way of distinguishing between the standard quantum interpretation of an entangled state and alternative explanations involving hidden variables. He based his ideas on the Bohm-Aharonov spin version of the EPR thought experiment. Bell's theorem would prove critical to further analysis of what is really happening when an observer measures a quantum system. It would be verified in 1982 through a polarization experiment carried out by French physicist Alain Aspect and his colleagues.

Bohm's and Bell's work pertained to the interpretation of quantum mechanics rather than to its applications. A more practical issue involved extending quantum field theory to include other forces besides electromagnetism. The goal was to generalize quantum electrodynamics to a theory that could describe other interactions such as the nuclear forces and gravity.

In that domain, around the time of Romer's "half hour with Einstein," a major theoretical breakthrough took place. In early 1954, physicist Chen-Ning "Frank" Yang and mathematician Robert Mills published a paper extending Weyl's gauge field theory to include other symmetry groups besides that of a simple circle. Recall that the original gauge theory, applied to electromagnetism, is something like a fan or weathervane that could point in any direction around a loop. Therefore, it has a kind of circular rotational symmetry.

The symmetry group of rotations around a circle is called U(1). A key property of U(1) is that it is Abelian, meaning that order of operations doesn't matter. If you spin a fan a quarter of the way around clockwise and then a third of the way around counterclockwise, you would reach exactly the same place if you reversed the order.

The work by Yang and Mills extended Weyl's method to non-Abelian symmetry groups. A simple example in nature is three-dimensional rotations, which can be represented by the group SU(2). Pick up an egg, mark a point on it carefully, and rotate it a quarter of the way clockwise around its longer axis and a third of the way counterclockwise around its shorter axis. Unlike the two-dimensional circle case, if you switched the order, the mark on the egg would reach a different position. In other words, order of operations does matter for non-Abelian groups such as SU(2).

An important aspect of Yang-Mills gauge theory (one that would later be proven by the Nobel Prize–winning work of Dutch physicists Gerardus 't Hooft and Martinus Veltman) is that, like QED, it is renormalizable, meaning that it produces finite answers in calculations. Its properties would turn out to be ideally suited for modeling the weak and strong nuclear interactions along with electromagnetism. Of course, Einstein would have little interest in a unification that included probabilistic aspects, such as one based on quantum field theory.

When Heisenberg stopped by Einstein's house in fall 1954 during a lecture tour of the United States, Einstein displayed just such a lack of interest. Over coffee and cake, Heisenberg tried one last time to persuade the founder of relativity about the probabilistic aspects of nature. He hoped to entice Einstein by mentioning that he had started working on his own unified field theory, based on quantum principles. To make the afternoon go smoother, they avoided all mention of politics. Nevertheless, Einstein was not impressed. Chiding Heisenberg, he kept repeating his old maxim, saying, "But you cannot believe, surely, that God plays at dice."[17]

A Pencil and Notepaper

After his meeting with Heisenberg, Einstein would live only about half a year longer. Since 1948, he had known that he had a ticking time bomb in his chest, an aortic aneurysm that could rupture at any moment. His fragile health was one factor that led him to limit his travel and spend most of his time in Princeton. He did travel down to Sarasota, Florida, once to rest up, but that was a rare jaunt out of town.

His sister Maja's death in 1951 had saddened him greatly. He felt more alone than ever before. One solace in his final years was that he grew closer to his son Hans Albert, who had moved to the United States and become a hydraulic engineering professor at Berkeley. Whenever

Hans Albert visited him, they made up for lost time by chatting about their mutual scientific interests.

Horrified by the prospect of nuclear war, Einstein spent much of his time campaigning for world government. Ceding control over weapons of mass destruction to a central global authority, he felt, would be the only way to prevent their use. Knowing that his time on earth was limited, he hoped to make his best case to preserve the planet.

A strong supporter of a Jewish homeland, he was dismayed by the raging conflict in the state of Israel, founded in 1948. Hoping that Jews and Arabs in the region could live together in peace and equality, he urged a negotiated solution to their land disputes. He wanted an Israel that was friendly with and accepted by its neighbors.

In 1952, when the first president of Israel, Chaim Weizmann, died, Einstein had been formally offered the country's presidency. Although greatly honored, he quickly and politely turned down the invitation. No doubt his heart condition and reluctance to travel played some part in his decision. The major factors were that he preferred solitude to being in the spotlight and that he had no interest in serving as head of state—especially if he came to disagree with the government's decisions.

Einstein's last major public act was signing the Russell-Einstein manifesto, a call for world peace initiated by philosopher Bertrand Russell. Arguing that the next world war would likely involve nuclear weapons, such as the hydrogen bomb, that could destroy major cities and threaten to annihilate the human race, the petition called for an end to armed conflict in favor of peaceful resolution of disputes. Einstein signed the document on April 11, 1955, only a week before his death.

Einstein's final few days were marked by intense pain. Nevertheless, he remained brave and alert. Dukas was startled on April 13 to find him collapsed on the floor. She called a doctor, who came over and prescribed morphine to help him rest. The next day, several physicians arrived and informed Dukas that Einstein's aneurysm had become unstable and would soon burst. They recommended surgery, but he refused, saying that he had lived long enough and it was time to go. When on the following day, he was immobilized with pain, Dukas called for an ambulance. He was taken to Princeton Hospital.

Even while in dire agony, Einstein still wanted to work on unified field theory. The day before he died, he asked for a pencil and his notes so that he could continue his calculations. His son had arrived, and was by his side throughout the day, along with his trusted executor Otto Nathan and Dukas.

In the early hours of April 18, Einstein's world line reached its final point—life's ultimate singularity. As the doctors had warned, the aneurysm suddenly burst. He muttered his last words in German to a nurse who didn't understand the language. Alas, they are lost to posterity.

Einstein never wanted a memorial or even a grave. Except for his brain, his body was cremated and the ashes scattered. Bizarrely, pathologist Thomas Harvey, while inspecting Einstein's cadaver before cremation, had made the unilateral decision to remove and preserve his brain for scientific research. In subsequent years, he sliced up parts of it and pieces were analyzed. Today, some of the slices are on display at the Mütter Museum in Philadelphia.

A fitting tribute to Einstein was held a few months after his death. Organized by Pauli, a major conference in Bern celebrated the jubilee of the special theory of relativity. It attracted leading researchers from around the world, including some, such as Bergmann, who were returning to Europe for the first time since the war. Touchingly, Einstein's last assistant, Bruria Kaufman, presented his final paper on unified field theory to the group.

Vienna Calling

With Einstein's death, Schrödinger lost one of his closest correspondents. Despite the debacle of 1947, they had still greatly trusted each other's opinions. It is fortunate that they had resumed correspondence before Einstein's passing; otherwise, Schrödinger would have had even deeper regrets.

Since 1946, Schrödinger had hoped to return to Austria. However, he was reluctant to return to Vienna when the city was partly occupied by Soviet troops and surrounded by the Soviet sector. Tired of politics, he had little desire to be a pawn in the Cold War. In his book, neutrality was the best policy.

He was delighted, therefore, when the former Allies reached an agreement in 1955 for the withdrawal of all foreign troops from Austria. In return, the country would solemnly promise to remain neutral and free of nuclear weapons indefinitely. From his perspective, having weathered Austro-Hungarian imperialism, Austrian fascism, and the Nazi Anschluss, it was the best political news of his lifetime. Offered a position at the University of Vienna, Schrödinger hoped for a creative post-Dublin career. As he and Anny boarded the ship to depart from his adopted city on the way back to his hometown, de Valera was the last to bid them farewell. It was a bittersweet moment, for as much as Schrödinger loved Ireland, he longed for the mountainous terrain of his native land. Upon arriving in Vienna, he was welcomed back by the Federal Ministry for Education. Austria was pleased to see the return of its illustrious countryman.

Sad to say, Schrödinger's homecoming was not as joyful and relaxing as he had anticipated. Erwin and Anny spent their final years in very poor health. Both had serious respiratory conditions. Along with bad asthma, Anny suffered from severe depression and had been receiving electroshock therapy. Before antidepressants, it was considered one of the standard treatments. Erwin had bouts of bronchitis and pneumonia, exacerbated by his lifelong smoking habit. Due to cataract surgery, he needed to wear thick glasses. He also developed phlebitis, atherosclerosis, high blood pressure, and a heart problem. When hiking, he often had to stop to catch his breath. He was frustrated not to be able to climb mountains he used to scale with ease.

Right before leaving Dublin, he had an episode of bronchitis so bad that, in his effort to get some rest, he overdosed on sleeping pills downed with whiskey. The next morning Anny found him virtually unconscious and had trouble waking him up. She called a doctor in a state of panic. Fortunately, the doctor was able to rouse him and he pulled through.

Once settled at the University of Vienna, Schrödinger tried to focus on his research. Despite his ailments, he managed to work on a few late-life projects. He mentored a young physicist, Leopold Halpern, who served as his last research assistant. Halpern would later work with Paul Dirac, the other 1933 Nobel laureate in physics.

Returning to the philosophical ruminations of his youth, Schrödinger wrote an essay, "What Is Real?," designed to complement his 1925 piece "Quest for the Path." He published the combined work as *My View of the World,* intended to be his definitive statement about the nature of life, consciousness, and reality. Several years earlier he had published a book about Greek philosophy, *Nature and the Greeks.*

In the spirit of Plato and Aristotle, Schrödinger always saw himself as more of a natural philosopher than an expert in calculations, although he was certainly adept at the latter too.

Transitions and Endings

On August 12, 1957, Erwin turned seventy. He soon decided that it was time to retire from the university. At the end of the academic year, he was awarded emeritus status, which offered him many of the amenities of being a professor but without the teaching duties. While it was unusual for someone to step down so soon after an academic appointment, Schrödinger had made a number of quick transitions in the past, particularly in his early career. Only the Dublin appointment had lasted for more than a decade.

There is no record of Schrödinger's reaction to a paper published in July 1957 by Princeton doctoral student Hugh Everett III, "'Relative State' Formulation of Quantum Mechanics." The paper detailed what later became known as the "many-worlds interpretation" of quantum mechanics, a clever alternative to the orthodox view. While the article is now considered a classic, few physicists read it at the time. Wheeler, Everett's doctoral advisor, encouraged his imaginative ideas but was concerned that mainstream physicists such as Bohr might find them outlandish. Indeed, Bohr was little interested in or impressed by Everett's work. Only after physicist Bryce DeWitt began to publicize the hypothesis in the 1970s would it begin to attract supporters.

Interestingly, Einstein had interacted with Everett much earlier. In 1943, when Hugh was twelve years old, he had written to Einstein and asked whether the universe was random or had a unifying principle. Einstein kindly wrote back, stating in effect that Hugh had created and vaulted over his own philosophical hurdle.

The many-worlds interpretation offers an unambiguous way of analyzing Schrödinger's cat scenario. It purports that each quantum observation would involve a branching of reality into myriad parallel paths. Everett cleverly addressed the question of determinism and the role of the observer by purporting that the observer's conscious existence would split seamlessly along with reality's branching. Therefore, each copy of the observer would believe that his or her scenario is the true, predetermined reality—and he or she would be right for that branch. No collapse would occur, eliminating the measurer's effect on what is measured. Consequently, placing a cat in a steel chamber with a radiation-triggered mechanism would lead to a bifurcation caused by the possibility of decay. In one branch the sample would decay, the cat poisoned, and the observer glum. In the other, the sample would be preserved, the cat spared, and the observer gleeful.

Everett came to believe that his interpretation implied immortality.[18] Given any agent that could cause death, there would always be a parallel branch in which survival would be possible. So if a cat was placed in the steel chamber an hour a day, one version of it would always live to see the next, and so on.

If such immortality were possible, we would not be aware of all the unfortunate copies of ourselves that met with cruel fates. We would not see the mourners in all the other parallel branches. We would, however, see our own loved ones pass away—at least so we think, from our branch's perspective. It is unclear, therefore, if such a type of immortality would be a blessing or a curse. There are echoes of this in the situation of Erwin and Anny in the late 1950s, by which time both had suffered through so many bouts of illness that each began to envision surviving without the other.

In 1958, Heisenberg made a belated entrance into the unification drama by publicly announcing his own unified field theory. Unlike Einstein's and Schrödinger's attempts, it was grounded in standard quantum mechanics and particle physics. Based on spinors (which are like vectors but transform differently), it incorporated what was known about the weak nuclear interaction, including the recent discovery by Yang and T. D. Lee that parity is not conserved. Parity conservation is the property that the mirror image of a process should be equivalent to the original process. As Yang and Lee had pointed out, processes involving the weak interaction, the force that explains many types of radioactive decay, do not always follow that rule. By then Schrödinger was out of the game and did not publicly comment on Heisenberg's unified theory—which lacked experimental evidence anyway.

The same year saw the death of Pauli, who had contributed to Heisenberg's unified theory. The physics world was stunned, as he was only fifty-eight and had been active at the time. He and Heisenberg had spent much of the year in a feud, which started when a press release called him "Heisenberg's assistant."[19] Insulted by the designation, he began to attack Heisenberg's theory openly. When he heard Heisenberg give a radio talk about his theory, stating that only the details needed to be filled in, Pauli sent physicist George Gamow a sketch of an empty rectangle with the inscription, "This is to show the world that I can paint like Titian. Only technical details are missing."[20]

Incensed with Pauli, Heisenberg shunned his funeral. It was a sad close to a once productive collaboration. Compared to Pauli and Heisenberg, Einstein and Schrödinger remained far more magnanimous, despite their brief battle in the press.

Two highlights of Schrödinger's final years were Ruth's wedding to Arnulf Braunizer in May 1956 and the birth of Ruth and Arnulf's first child, Andreas, in February 1957. Several years earlier, Erwin had revealed to Ruth that he was her biological father. Therefore, he could openly revel in being a grandfather. Unfortunately, Ruth's legal father, Arthur March, passed away shortly after Andreas was born.

The Braunizers settled in Alpbach, a delightful Tyrolean mountain village near Innsbruck. With its fresh air and abundance of flowers, it was a favorite place for the Schrödingers too. It offered a respite from busy Vienna, and they found there much rest and comfort. As the time of this writing, Ruth and Arnulf are still living there.

In May 1960, Erwin received dire news from his physician: the tuberculosis that he hoped he had conquered decades earlier had reappeared. As the year progressed, his breathing became more and more labored, until he was admitted to a hospital, where he spent the Christmas holidays.

He had let Anny know that he wished to spend his last moments at home, not in a clinical setting. Released from the hospital, she took him home and stayed by his side, gently grasping his hand. The challenges of their waning years had brought out the deep affection they still held for each other. His last words were a call of devotion to her.

On January 4, 1961, Schrödinger departed from the material world. Supervised by Hans Thirring, his body was taken to a coroner for postmortem examination and then transported to Alpbach, where he was buried in a churchyard on January 10. Thirring delivered the eulogy for his longtime friend. The grave was marked by a wrought iron cross, superimposed with a circle on which his famous wave equation was etched.

Many years later, Ruth placed a plaque with one of Schrödinger's poems in front of the grave marker. Including the line "all being is one being," it nicely sums up his Vedantic philosophy that everything is interconnected and eternal.[21] The blend of poetry in the plaque and physics in the marker honor his complex soul just perfectly.

A Cat Creeps into Culture

At the time of Schrödinger's death, physicists knew him primarily for his wave equation, while biologists (and biology enthusiasts) were familiar with him mainly because of *What Is Life?* However, the general public was largely unaware of his cat paradox, the contribution that ended up becoming his most famous. That changed in the 1970s when several works of science fiction brought attention to his entangled story.

One of the first speculative tales about the subject, "Schrödinger's Cat" by Ursula K. Le Guin, appeared in 1974. Le Guin had learned about the quantum thought experiment by reading "physics for peasants," as she put it. "It was obviously an excellent metaphor for a certain kind of science fiction."[22]

Other writers in subsequent years have sired a litter of whimsical quantum cat tales. Many of these have focused on parallel universes and related themes. In 1979, Robert Anton Wilson published *The Universe Next Door*, the first book in the Schrödinger's Cat trilogy about alternative histories. Robert Heinlein's *The Cat Who Walks Through Walls*, published in 1985, envisioned new realities spawned by time travel. Around that time, several popular science books discussed the implications of the paradox. A menagerie of quantum animal stories has followed—typically involving cats, but sometimes relating to other creatures or even people trapped in ambiguous circumstances between life and death.

A poem published in 1982 by writer Cecil Adams in his column "The Straight Dope" has become part of quantum kitty lore (especially since it became widely available on the Internet much later). It describes an epic battle between Win (Schrödinger) and Al (Einstein) about chance in the universe that engenders the cat paradox and the dice-rolling remark. The saga ends with Win laying bets at Al's funeral over whether he'd make it to heaven.

After clawing its way through literature, the eerie cat leapt into the world of pop music, thanks to the band Tears for Fears. The group released the song "Schrödinger's Cat" as the B-side of a single in the early 1990s. (Later they also released "God's Mistake" with the lyric "God does not play dice"—transforming Einstein's statement into a musing on love's unpredictability.) As songwriter Roland Orzabal explained: "My song . . . is merely a dig at the classical scientific way of seeing things, a dig at rational materialism, at taking things apart without being able to put them back together, at seeing the trees and not the wood. At the end of the song, I sing, 'Schrödinger's Cat is dead to the world.' Is the cat dead, or just asleep? I like the ambiguity, the uncertainty."[23]

In recent years Schrödinger's cat has become a popular meme. It has appeared on T-shirts, in cartoons (such the popular online comic strip *xkcd*), and in television shows (such as *The Big Bang Theory* and *Futurama*). In perhaps its most prominent mention, Google incorporated a doodle of the experiment into its logo on August 12, 2013, the 126th anniversary of Schrödinger's birth. Through these varied cultural references the cat—and even the term "Schrödinger's" applied to anything—has come to be a symbol for ambiguity in general.

Scientific Legacies

Much of what we know about the complex lives of Schrödinger and Einstein has been revealed through archival materials. Unfortunately, the value of their intellectual estates has meant a prolonged series of battles for control.

In 1963, Anny received a visitor from the United States, philosopher and historian of science Thomas Kuhn. Kuhn had become part of a project to document the history of quantum physics. After sitting for an interview, Anny offered Kuhn a large box, weighing more than two hundred pounds, that was full of her late husband's letters, manuscripts, diaries, and other personal materials. It was an unmatched treasure trove of Schrödinger material that would prove invaluable to historians.

Kuhn carefully arranged for much of the material to be duplicated (mainly on microfilm) and donated the originals to the University of Vienna Central Library. The library has preserved the box for decades, while researchers have perused the copies in repositories and research centers all over the world.

After Anny died in 1965, Ruth became the sole heir of Schrödinger's estate, but she did not learn about the box until the 1980s. She spoke with Walter Thirring, head of the Physical Institute at the university, but was told that no new materials were available. In 2006, she asked the university rector for the return of the materials. The university sought legal counsel and decided to sue to settle rights of possession. The Braunizers hired their own lawyer, and legal wrangling began in an effort to establish who had proper ownership of the materials.[24]

The case has dragged on for a number of years. Much progress was made in autumn 2008 when both sides agreed on the steps for a possible solution.[25] The idea was to establish a new foundation to administer the materials. Finally, in October 2014 the case was settled and Schrödinger's papers designated a UNESCO World Heritage Site.

Einstein's papers were also the subject of a legal skirmish. After he died, Otto Nathan and Helen Dukas administered his estate. They personally approved use of his image and materials until the bulk of the materials were transferred to Hebrew University, Jerusalem. A duplicate archive was established in Princeton, enabling researchers to have access to his papers. Nathan and Dukas signed an agreement with Princeton University Press for them to begin publishing his writings in edited volumes. However, in the 1970s a dispute arose between Nathan and the press about its choice of editors, and a court needed to step in and arbitrate. Physicist and science historian John Stachel became lead editor of the project.

Then came a turn of events no one had anticipated. Stachel and another historian, Robert Schulmann, learned of a safe-deposit box in Berkeley where Hans Albert's second wife, Elizabeth, had stored a cache of some five hundred letters between Einstein and Mileva. The collection included about fifty early love letters that shed light on a hitherto unknown period of Einstein's life. After further disputes between Einstein's estate and Princeton University Press, the press obtained the rights to publish the love letters. Many readers were stunned by the contrast between the passion Albert expressed for Mileva at the start of their relationship and the disdain he expressed later on before their divorce.

The lives of Einstein and Schrödinger inform us that even the most brilliant scientists are human. Along with spectacular bursts of insight, they endure long intervals in which their wheels grind away without offering any traction. In their personal relationships, they have times of tenderness and incidents of betrayal. They might race after fleeting illusions and then come running home to those who really care about them.

The correspondence between Einstein and Schrödinger conveys considerable warmth and mutual support. Perhaps, like Don Quixote and Sancho Panza, they had ended up chasing windmills. They each knew that their quests could be dismissed as quixotic, their lives seen as eccentric. Yet the *compañeros* stuck by each other—if not always in the glare of the press, then in the depths of their hearts.

Beyond Einstein and Schrödinger:
The Ongoing Search for Unity

With photography, at least there is one good thing,
once you take your picture you are through with it. It is
finished. But with a theory, it is never finished.

—ALBERT EINSTEIN, reported in the *Christian Science Monitor,*
December 14, 1940

Who will be the next Einstein? Will his ingenious contributions ever be surpassed? Is there anyone brilliant enough to complete his dream of a unified theory of nature? We have seen that despite being an accomplished physicist, Nobel laureate, and Renaissance man, Schrödinger never came close to Einstein's fame internationally (beyond Ireland in the 1940s, that is). If anything, his cat has taken all the glory—at least as a cultural meme. However, he certainly hasn't been the only one to try to fill Einstein's shoes.

Since 1919, when the public first tasted the theory of relativity through the announcement of the solar eclipse measurements, it has had an insatiable appetite for news about Einstein and possible successors. While he was alive, as we've seen, the press trumpeted every unified field theory he proposed as if it were a major breakthrough. After his death, stories about brilliant individuals tantalizingly close to completing his mission have continued to make headlines. All in all, Einstein, his unfinished quest, and the question of who might inherit his throne have served as touchstones for almost a century.

Research scientists know that progress in any field is usually incremental, taking place over years or even decades. Groundbreaking discoveries are few and far between. Often a scientist needs to be lucky enough to be in the right place at the right time to make a mark. Most scientific research today is completed by large teams, rather than by single individuals.

Yet the myth persists of the lone genius changing everything around us. Type "next Einstein" into any Internet search engine and expect to be bombarded with results—everything from recipes for educational success to claims made in resumes or personal ads. Here are a few assorted examples of recent media musings: Will the next Einstein be a "surfer dude"?[1] Is he a child prodigy with an exceptional IQ?[2] What if the next Einstein is a computer?[3] Could a smartphone application identify him?[4] Or might an old-fashioned DVD designed for little ones do the trick? As a 2009 *New York Times* headline advised with tongue firmly in cheek, "No Einstein in Your Crib? Get a Refund!"[5]

The formula that produced Einstein was a perfect match between crucial scientific problems that demanded radical solutions, exceptional insights that often overturned commonly held beliefs, an ironically photogenic visage (who knew that rumpled sweaters, a Brillo pad mustache, and a mop of unruly gray hair could be so compelling?), and the omnipresent glare of the camera. His rise to fame coincided, more or less, with the golden age of Hollywood, when cinematic newsreels projected the latest fashions, feats, and foibles of celebrities. Like Douglas Fairbanks, Mary Pickford, Charlie Chaplin, the Barrymores, and countless other stars of the cinema in the twenties, thirties, and forties, Einstein traipsed across the screens of thousands of Main Street movie houses. The public viewed him stopping on his strolls to wave to admirers, giving speeches about current affairs, headlining benefits for various charities, and occasionally reporting progress in his research. Hungry to fill their quota of human interest stories, reporters lapped up news about the German Jewish scientist like scrawny cats with spilled milk.

It is not clear if that formula will ever be repeated. For one thing, there has been an explosion of publications. Many theories vie for prominence—far more than in the days of Einstein and Schrödinger. Yet the energies required to test these approaches have required increasingly expensive and time-consuming projects, such as the Large Hadron Collider near Geneva, Switzerland. Unlike, for example, the eclipse measurements, experimental science has generally proceeded far more slowly and cautiously, requiring far greater quantities of data before announcing results. In high-energy physics, teams typically involve hundreds of researchers rather than lone pioneers. At the same time, the media have diversified, so not everyone's eyes are fixed on the same scientific celebrities.

Peter Higgs, one of the recipients of the 2013 Nobel Prize in physics, has become a rare contemporary example of a well-known, accomplished theorist. Yet his name recognition hardly rivals Einstein's. The particle named after him, the Higgs boson, has come to be known colloquially as the "God particle." When it was discovered in 2012, much of his press coverage was shared with a divine being. (To India's dismay, its native son Satyendra Bose hardly got any mention.)

Triumph of the Standard Model

The discovery of the Higgs boson has supplied the last missing puzzle piece of the standard model of particle physics—the closest thing we have today to a unified field theory. The standard model includes a unified explanation of electromagnetism and the weak interaction, in tandem known as the electroweak interaction. It also contains a description of the strong interaction, the force that cements protons and neutrons together into atomic nuclei. The odd force out is gravity, which is not part of the standard model.

The development of electroweak unification began in 1961, the same year as Schrödinger's death, when physicist Sheldon Glashow suggested that electromagnetism and the weak interaction could be united through a theory involving four exchange (force-conveying) bosons: the photon, two charged bosons, called the W^+ and W^-, representing weak decay, and a fourth boson, later called the Z^0, representing a weak neutral exchange. At that point the fourth type of interaction, between two particles of similar charge, hadn't yet been observed. The Lagrangian (description of energy) Glashow used wasn't quite correct, but his idea of four exchange particles was right on the mark.

A vexing problem with uniting electromagnetism with the weak force, however, was that the two forces have vastly different ranges and interaction strengths. Electromagnetism acts over an enormous range, which we witness when we observe the light from stars that are trillions of miles away. The weak force, in contrast, operates only on the nuclear scale. Furthermore, on the subatomic level, electromagnetism is about ten million times stronger than the weak force. If in the early days of the universe those forces were united, why do they seem so different today?

As physicists came to realize, it is the properties of the exchange bosons, volleying back and forth between material particles, that determines the ranges and strengths of forces. Massless bosons, such as photons, create potent, long-range forces. Heavy bosons, such as the W and Z exchange particles, generate weaker, short-range forces. Consequently, explaining the present-day discrepancy between the electromagnetic and weak interactions boils down to understanding how the W and Z bosons acquired mass.

Enter the Higgs mechanism, a brilliant way of understanding how, as the universe cooled down from its fiery Big Bang beginning, most types of particles acquired mass, while photons didn't. Proposed independently in 1964 by several groups of researchers, including Higgs, François Englert (who was awarded the Nobel along with Higgs) and Robert Brout, and a team consisting of Gerald Guralnik, Carl Richard Hagen, and Thomas Kibble, it imagines that a field with a certain type of gauge symmetry pervaded the early universe. The spontaneous breaking of that symmetry that accompanied the lowering of space's temperature endowed most particles with mass, leaving the photons massless.

We imagine that gauge symmetry as a kind of whirling fan placed at each point in the field, spinning around and puffing out air in every direction. As the universe cooled, its conditions became such that the initial symmetry of the Higgs field became spontaneously broken. Each fan froze into place, with all of them aimed in the same direction. Before the freezing up, the spinning actions of the fans canceled each other out, allowing all particles to move freely any way they pleased. However, once the fans froze in place and blew air at identical angles, the blowing hindered most particles, shortening their range and reducing their strength. In other words, they acquired mass. Only photons, which did not interact with the blowing air, remained massless. They kept their full strength and long-distance range.

In the late 1960s, American physicist Steven Weinberg and Pakistani physicist Abdus Salam independently constructed Lagrangians (along the lines of the Yang-Mills gauge theory, described earlier) that included Higgs field components along with exchange bosons and fermion fields representing matter particles. Their Lagrangian was designed to undergo spontaneous symmetry breaking below a certain temperature, at which point three of the exchange bosons, the W^+, W^-, and Z^0, would acquire mass via the Higgs mechanism, leaving the photons massless. The fermions would also accumulate mass. A segment of the original Higgs field would remain as a massive particle called the Higgs boson.

By then, so many new elementary particles had been discovered that choosing which fermions to label as fundamental proved critical. Most physicists suspected that protons and neutrons were not fundamental but instead were made up of constituents. The subcomponents were at first called different things, but eventually the physics community settled on the term "quarks," chosen by Murray Gell-Mann for the way it sounded to him. He spotted the word in a passage from James Joyce's *Finnegans Wake,* "Three quarks for Muster Mark." As there are three quarks each in protons and neutrons (and in all particles in the category called baryons), the moniker seemed appropriate.

Once quarks were catalogued, they seemed to fall into distinct families, called generations. The first generation, including "up" and "down," comprises the quarks that form protons and neutrons. The second generation, called "strange" and "charmed," includes more massive, exotic particles. Finally the third, even heavier generation, called "top" and "bottom," was not discovered until the 1980s (bottom) and 1990s (top). Each generation also includes antimatter particles of the same mass but opposite charge, called antiquarks. The specific types of quarks, such as "up" and "strange," are called "flavors."

Leptons, particles that don't experience the strong force, similarly fall into three generations. The first consists of electrons and neutrinos: extremely light, fast-moving particles. The second includes muons and muon neutrinos. Massive tauons and tau neutrinos make up the third category.

Unlike the unification proposals of Einstein and Schrödinger, electroweak unification theory offered many specific, testable predictions. These included the presence of a weak neutral current (weak interaction between particles of like charge), the existence of the W^+, W^-, and Z^0 exchange bosons at certain masses, and the actuality of the Higgs boson. Over the course of the 1970s and 1980s, particle accelerator experiments at CERN (European Organization for Nuclear Research), near Geneva, Switzerland, verified each of those predictions except for the last. Finally, the Higgs boson was confirmed through particle collision data collected at CERN's Large Hadron Collider.

Along with electroweak unification, the standard model also includes a theoretical description of the strong interaction that involves exchange particles called gluons. These form the "glue" that sticks quarks together and keeps them confined to groups of three (or quark-antiquark pairs in the case of mesons). In an analogy to positive and negative electric charges, each quark has a color charge. "Color," in this context, has nothing to do with visual appearance; it is just shorthand for a particular conserved quantity. By volleying gluons between quarks of different colors, the strong force naturally emerges. The quantum field theory describing this is called quantum chromodynamics (QCD), in an analogy to quantum electrodynamics.

Given how the standard model has shaped up, it is amusing to think of all the newspaper pronouncements proclaiming Einstein's and Schrödinger's unified field theory proposals as ultimate descriptions of the universe. The picture of nature that has emerged in recent decades is radically different from what anyone in the World War II era anticipated. Clearly the universe has many surprises up its sleeve. Could it be that new discoveries will someday make the standard model seem outdated?

Mind the Gaps

Over the years, the predictions of the standard model have been tested again and again to extraordinarily high precision. By that measure, it is a remarkably successful theory, explaining everything from kitchen magnets to the Sun's dynamo. It offers an unprecedented type of unification that encompasses three out of the four forces of nature. Only gravity is left out.

The same level of certainty applies to general relativity. Numerous high-precision experiments have verified the many predictions of Einstein's masterly theory of gravitation. Recent tests include satellite measurements of a phenomenon called "frame dragging," proposed by Schrödinger's old friend Hans Thirring and fellow Austrian physicist Josef Lense back in 1918. Frame dragging involves the distortion of space and time around Earth because of its rotation. In February 2016, researchers announced the discovery of gravitational waves, confirming another major prediction of general relativity.

Combine the standard model with general relativity and you have a powerful tool kit for exploring nature's properties. But is that enough? Not if you look at the glaring omissions that neither theory can explain. Dark energy, the agent for accelerating the universe, and dark matter, the invisible substance that keeps galaxies from flying apart, represent mysteries on par with those that challenged the quantum pioneers. We have mentioned how the former seems to match the cosmological constant term proposed (and later retracted) by Einstein and later advocated by Schrödinger. However, no one knows the physical source of dark energy, which acts as a kind of antigravity.

The nature of dark matter offers another modern-day conundrum. First identified in the 1930s by Swiss astronomer Fritz Zwicky in his study of the Coma cluster, it constitutes the unseen mass gravitationally required to keep astronomical structures stable. As Zwicky's claim was not taken seriously, it took another half century before the search for dark matter began in earnest. The trigger was the finding by astronomers Vera Rubin and Kent Ford that Andromeda and other galaxies don't have enough visible matter to keep their outer stars moving as fast as they actually do. Galaxies seem to act like merry-go-rounds, with speedy outer horses pulled by unseen mechanisms. Starting in the 1980s, astronomers and particles physicists have conducted searches for dim astronomical objects and/or invisible particles with enough gravitational oomph to constitute dark matter. The focus began to center on cold (slow-moving) dark matter particles that respond to the weak force and gravity but not to electromagnetism (hence their invisibility). Searches for such particles have been conducted in converted mine tunnels deep underground, to avoid the "noise" of ordinary particles, as well as in space. At the time of this writing, conclusive evidence for dark matter particles has yet to be found.

If dark energy and dark matter were rare phenomena, perhaps we could hold off on explanations and try to tidy up other loose ends in physics. On the contrary, together they constitute 95 percent of everything in space. According to recent astronomical estimates, a whopping 68 percent of the universe is dark energy and fully 27 percent is dark matter, leaving only 5 percent that can be explained through the standard model combined with conventional general relativity. Some have suggested modifying general relativity, continuing along Einstein's path to enhance it. However, the bulk of the physics community recognizes the overwhelming success of both the standard model and general relativity in describing what we actually can observe. The desire not to tamper with success leads to the quandary of how to move beyond, and perhaps even unify, those two twentieth-century masterworks.

Aside from the question of the universe's dark substances, other mysteries remain for the standard model. Why do some particles (quarks) feel the strong force, while others (leptons) don't? Can science explain why there is so much more matter than antimatter in the observable universe? Why are there only three generations of constituents and why do they have particular masses? Is there a way of interchanging fermions and bosons that offers a link between matter particles and energy fields? Those are among the many open questions in particle physics today.

Dreams of Geometry, Symmetry, and Unity

In recent decades the dream of Einstein, Schrödinger, Eddington, Hilbert, and others to explain everything in the cosmos through pure geometry has undergone a marked revival. It seems that every time science veers away from the Pythagorean ideal that "all is number," abstract thinkers strive for ways of steering it back.

Instead of theoreticians imagining matter waves (of the de Broglie/Schrödinger type) oscillating on the atomic level, many now envision strings (filaments) and membranes (surfaces) of energy vibrating at much more minuscule scales. These strings and membranes are purely geometric structures that, through their twisting and jiggling, generate the known particle properties. String theory is a vast subject; let's take a brief look.

The initial impetus for string theory was an unsuccessful attempt by Japanese physicist Yoichiro Nambu and others in the late 1960s and early 1970s (before the gluon idea took hold) to model the strong interaction by connecting particles together via flexible, energetic strands. Such "bosonic strings," as they were called, acted like a dog's leash, confining a particle to a tiny region (nuclear scale) while offering freedom within those bounds.

In 1971 French physicist Pierre Ramond discovered a way of modeling fermions as strings as well. He developed a method, called "supersymmetry," in which bosonic strings could be transformed into fermionic strings by a kind of "rotation" through an abstract space. His breakthrough inspired theoreticians John Schwarz and André Neveu to develop a comprehensive theory of both building-block fermions and force-wielding bosons using strings vibrating in different manners. Such all-purpose strands were dubbed "superstrings." One peculiar aspect of superstring theory is that it is mathematically complete (lacking terms seen as unphysical) only in a space of ten dimensions or more. Earlier that year, physicist Claud Lovelace had shown that bosonic strings require twenty-six dimensions, so needing only ten seemed like an improvement.

By the mid-1970s, physicists were cracking open textbooks and articles that described the Kaluza-Klein theory in higher dimensions, hoping to learn how to handle them. A primer Bergmann wrote about general relativity in the 1940s, with an introduction by Einstein, helped reacquaint the theoretical community with methods for dealing with more than four dimensions. Oskar Klein's old idea of compactification—wrapping up extra dimensions so tightly that they cannot be observed—underwent a revival. Theorists found ways of looping the six additional dimensions around tiny, tightly bound spaces, like minute balls of twine. Mathematicians Eugenio Calabi and Shing-Tung Yau would develop a classification scheme for such twisted-up spaces, called Calabi-Yau manifolds.

Excitement built among the physics community in 1975 when Schwarz and French physicist Joël Scherk proposed a way of explaining gravity using supersymmetry. They showed how gravitons, the hypothesized bosons conveying gravitational attraction, would naturally emerge in their theories by applying methods of supersymmetry to other types of particles. Gravitation, the researchers argued, was thereby the natural consequence of a union between bosons and fermions. Wed the two types and gravitons are born.

Some researchers, particularly French theorists Eugène Cremmer, Bernard Julia, and Scherk, working at the École Normale Supérieure in Paris, Dutch physicist Bernard de Wit, working with German physicist Hermann Nicolai, Dutch physicist Peter van Nieuwenhuizen's group at Stony Brook, and others, applied supersymmetry to a standard (one not utilizing strings) quantum field theory, in an approach called supergravity. Cremmer, Julia, and Scherk showed how such a theory could ideally be housed in an eleven-dimensional spacetime, with seven of the dimensions compact. Despite initial promise, supergravity ran into problems representing certain aspects of the particle world.

Working with British physicist Michael Green, Schwarz continued to explore the properties of superstrings. In 1984, Green and Schwarz announced that they had developed a ten-dimensional model that was free of anomalies (mathematical bugs). Moreover, unlike QED, electroweak theory, and other standard quantum field theories, superstring field theory produced finite values and therefore did not require the canceling of infinite terms through renormalization. Their results, dubbed the "superstring revolution," offered much cause for celebration. Perhaps through superstrings, many physicists thought, Einstein's quest for unity could finally be fulfilled.

Just as Einstein, Schrödinger, and others had discovered that there were many ways to extend general relativity, Green, Schwarz, and other researchers—such as brilliant theorist Edward Witten of the Institute for Advanced Study, who proved key theorems—came to appreciate the many types of superstring theory. In fact, there was an embarrassment of riches. Superstring theory soon became a labyrinth with myriad possible routes. Who would supply the Ariadne's thread leading to a single, comprehensive theory of nature?

At a 1995 conference in California, Witten declared a second superstring revolution, this time involving supplementing strings with membranes. He dubbed the new approach "M-theory," enigmatically saying that while the "M" could signify "membrane," it could also stand for "magical" or "mystery." M-theory united several different kinds of string theory, along with supergravity, into a single methodology. One innovation, explored in the late 1990s by Nima Arkani-Hamed, Savas Dimopoulos, Gia Dvali, Lisa Randall, Raman Sundrum, and others, was the idea that one of the extra dimensions could be "large" (meaning not microscopic) but inaccessible to all types of fields except gravitons. That would explain why gravity is much weaker than the other natural forces.

Unlike the standard model and general relativity, nary a shred of evidence has turned up in support of supersymmetry, superstring theory, M-theory, or extra dimensions. Why, then, do these ideas have so much backing among theorists? Factors such as mathematical beauty, symmetry, and completeness—strikingly similar to some of Einstein's criteria—all come into play. Plus there aren't many other credible alternatives.

Loop quantum gravity, developed by Abhay Ashtekar, Carlo Rovelli, Lee Smolin, and others, offers perhaps the mostly widely supported method for quantizing gravity other than strings. Like Schrödinger's general unitary theory, it emphasizes the primary role of the affine connection, which is modified and used as a quantum variable. Spacetime becomes replaced with a kind of geometric foam. String theorists often point out that loop quantum gravity does not offer a theory of everything, just a quantized theory of gravity. Loop quantum gravity supporters argue back that string theory treats gravitation both as a background (the spacetime metric in which fields move) and as a field (gravitons) rather than as a unified whole. Their goal is to understand quantum gravity first, before trying to wed it to other interactions.

Exploring the full implications of string/M-theory and loop quantum gravity would require an excursion to the Planck scale, the minuscule domain in which quantum theory and gravity meet. Such tremendous energies are well beyond current reach. Fortunately, high-energy theories often have lower-energy implications. Thus the Large Hadron Collider could well detect particle states that offer a window into physics beyond the standard model. An example would be supersymmetric companion particles: mates of fermions with bosonic properties, or vice versa. The discovery of such would offer powerful evidence for supersymmetry and possible dark matter candidates. While none have turned up so far, many physicists continue to be hopeful that superpartners will emerge in collider data once enough of it is collected and analyzed.

Faster Than Light: A Cautionary Tale

Researchers, students, funding agencies, science aficionados, writers, and others interested in what lies beyond the standard model are waiting eagerly for the tiniest hint of new, unexplained phenomena. With so much time and money invested in the Large Hadron Collider and other big-science experiments, no wonder there is much anticipation of groundbreaking results.

Physicists need to be careful, though, not to offer hasty announcements of success, no matter how tempting. The teams that identified the Higgs boson waited patiently for statistics to rule out other possibilities, even if that process took many months. They offered a lesson in perseverance. However, there are sometimes cases of researchers jumping the gun—making claims before other groups supply critical corroboration.

Although the Einstein-Schrödinger debacle took place in the 1940s, its lessons ring true today. Tight funding often requires scientists to justify the importance of their research, typically through press releases. A premature announcement of an unverified finding can leave a lasting impression that taints future research in that area. Even if that claim is refuted, the public might long remember it as an actual breakthrough rather than as a false report.

Take, for example, a research group's claim in September 2011 that they had detected faster-than-light particles at a facility in Gran Sasso, Italy. While much of the scientific community was dubious or at least cautious, the assertion was widely reported in the international press. A debate began in the media about whether Einstein's special theory of relativity needed to be modified. Reports wondered if the results would open the door to new physics beyond the standard model. Overlooking decades of experiments confirming special relativity and its light-speed limit, the claim was presented as a litmus test for relativity and the sanctity of the law that every effect must be preceded by its cause. For example, a piece in the British newspaper the *Guardian* reported, "Scientists at the Gran Sasso facility will unveil evidence . . . that raises the troubling possibility of a way to send information back in time, blurring the line between past and present and wreaking havoc with the fundamental principle of cause and effect."[6]

The alleged evidence for faster-than-light travel was put forth by a group called OPERA (Oscillation Project with Emulsion-tRacking Apparatus) that tracked streams of neutrinos emanating from the CERN accelerator laboratory near Geneva, Switzerland, about 450 miles away. After a three-year run, the team measured the neutrinos' times of arrival to be approximately sixty billionths of a second earlier than they would have been at light speed—assuming that their experimental apparatus was accurate.

"This result comes as a complete surprise," announced OPERA spokesperson Antonio Ereditato in a press release. "After many months of studies and cross checks we have not found any instrumental effect that could explain the result of the measurement."[7]

The press release and news reports emphasized that the findings required independent verification and should not be taken at face value. However, the monumental implications of such a discovery soon had the Internet—including the Twitter social network—abuzz with speculations and corny jokes.

As the *Los Angeles Times* headlined just days after the announcement, "Neutrino Jokes Hit Twittersphere Faster Than the Speed of Light." The accompanying article included examples such as: "We don't allow faster than light neutrinos in here, said the bartender. A neutrino walks into a bar."[8]

Songwriters soon joined in on the craze, including an Irish band, Corrigan Brothers and Pete Creighton, with their "Neutrino Song." "Was old Albert wrong?" asked one of the verses. "That fabulous theory of relativity is being debunked."[9]

If Einstein's theory had been shattered, theoretical physics would have faced a unexpected challenge. Perhaps it would have taken a new Einstein to pick up the pieces and assemble a more durable theory. But as has often been the case, reports of the demise of relativity were greatly exaggerated.

In June 2012, CERN issued an Emily Litella–esque "never mind" with a press release stating that "the original OPERA measurement can be attributed to a faulty element of the experiment's fibre optic timing system." Neutrino velocities, as confirmed by OPERA and three other experiments, do not exceed the speed of light. That is "what we all expected deep down," stated CERN research director Sergio Bertolucci.[10]

By the time of the curtain call for the OPERA episode, the faster-than-light neutrino meme had long faded from the Twittersphere and other media. Yet, without a doubt, the original announcement had generated unnecessary public confusion about science. On the Google search engine, for example, queries about neutrinos have continued to suggest "faster than light" as a common related expression. Who knows how many students doing research for essays have encountered some of the early reports in their search results and mentioned faster-than-light particles as a distinct possibility?

As a result of the incident, Ereditato and OPERA physics coordinator Dario Autiero decided to step down after a majority of the team voted against them in a no-confidence ballot. The vote and their subsequent resignations reflected a feeling that the leaders' announcements had been far too premature.

The Road Ahead

Patience is not a hallmark of the press, especially in the Internet age of rapid-fire news reports. The media hungrily gobble content as long as they can make the case that it is new and interesting to the public. Unpublished reports, speculations, preliminary results, and other findings not yet verified through the scientific review process sometimes become as newsworthy as meticulously verified conclusions.

Patience is also not an attribute politicians are often known to have, particularly during election years. We see how de Valera's political fortune depended, to some extent, on whether the Dublin Institute for Advanced Studies and his other pet projects turned out to be smashing successes or money-draining boondoggles. That knowledge drove Schrödinger—and de Valera's mouthpiece, the *Irish Press*—to trumpet his preliminary calculations as if they were the sacred tablets handed down from Mount Sinai. Schrödinger leapt to announcements only weeks after he completed his mathematical manipulations, and well before any review process checked them over. In modern times, science budgets have become easy targets, placing added pressure on researchers to proclaim their achievements.

Yet patience is precisely the quality needed during what looks like a long slog ahead for fundamental physics to reach its next milestones. Who knows when the first evidence will be found of phenomena transcending the standard model? What will be the cost of success? How many years of data collection and statistical analysis will be necessary before proof positive of new physics will be established?

We've seen the hazards of hasty reporting that downplays the time-tested process of verification. It confuses the public and ultimately doesn't help scientists. While both Einstein and Schrödinger fell victim at times to a mixture of wishful thinking and unwarranted publicity for their highly speculative unification hypotheses, in their quieter moments they emphasized the need for a deep, reflective, sober-minded reading of science. We would do well to read their writings, as well as those of the scientists and philosophers who inspired them, to contemplate the current state of physics and where to go from here.

Further Reading

(Technical works are marked with an asterisk.)

Aczel, Amir, *Present at the Creation: Discovering the Higgs Boson* (New York: Random House, 2010).

Cassidy, David C., *Beyond Uncertainty: Heisenberg, Quantum Physics, and the Bomb* (New York: Bellevue Literary Press, 2010).

———, *Einstein and Our World* (Amherst, NY: Humanity Books, 2004).

Clark, Ronald W., *Einstein: The Life and Times* (New York: Avon Books, 1971).

Crease, Robert P., and Charles C. Mann, *The Second Creation: Makers of the Revolution in Twentieth-Century Physics* (New Brunswick, NJ: Rutgers University Press, 1996).

Davies, Paul, *Superforce: The Search for a Grand Unified Theory of Nature* (New York: Simon and Schuster, 1984).

Einstein, Albert, *Autobiographical Notes,* translated and edited by Paul Arthur Schilpp (La Salle, IL: Open Court, 1979).

———, *Ideas and Opinions,* translated by Sonja Bargmann (New York: Bonanza Books, 1954).

———, *The Meaning of Relativity* (Princeton: Princeton University Press, 1956).

———, *Out of My Later Years* (New York: Citadel Press, 2000).

*Einstein, Albert, and Peter Bergmann, "On a Generalization of Kaluza's Theory of Electricity," *Annals of Mathematics* 39 (1938): 683–701.

Farmelo, Graham, *Churchill's Bomb: How the United States Overtook Britain in the First Nuclear Arms Race* (New York: Basic Books, 2013).

———, *The Strangest Man: The Hidden Life of Paul Dirac, Mystic of the Atom* (New York: Basic Books, 2009).

Fine, Arthur, *The Shaky Game: Einstein, Realism and the Quantum Theory* (Chicago: University of Chicago Press, 1986).

Fölsing, Albrecht, *Albert Einstein: A Biography,* translated by Ewald Osers (New York: Penguin, 1997).

Frank, Philipp, *Einstein: His Life and Times* (New York: 1949).

Freund, Peter, *A Passion for Discovery* (Hackensack, NJ: World Scientific, 2007).

Gefter, Amanda, *Trespassing on Einstein's Lawn: A Father, a Daughter, the Meaning of Nothing, and the Beginning of Everything* (New York: Bantam, 2014).

Goenner, Hubert, "Unified Field Theories: From Eddington and Einstein up to Now," in *Proceedings of the Sir Arthur Eddington Centenary Symposium*, edited by V. de Sabbata and T. M. Karade, 1:176–196 (Singapore: World Scientific, 1984).

Greene, Brian, *Fabric of the Cosmos: Space, Time and the Texture of Reality* (New York: Vintage, 2005).

Gribbin, John, *Erwin Schrödinger and the Quantum Revolution* (Hoboken, NJ: Wiley, 2013).

———, *In Search of Schrödinger's Cat: Quantum Physics and Reality* (New York: Bantam, 1984).

———, *Schrödinger's Kittens and the Search for Reality: Solving the Quantum Mysteries* (New York: Little, Brown, 1995).

Halpern, Paul, *Collider: The Search for the World's Smallest Particles* (Hoboken, NJ: Wiley, 2009).

———, *Edge of the Universe: A Voyage to the Cosmic Horizon and Beyond* (Hoboken, NJ: Wiley, 2012).

———, *The Great Beyond: Higher Dimensions, Parallel Universes, and the Extraordinary Search for a Theory of Everything* (Hoboken, NJ: Wiley, 2004).

Henderson, Linda Dalrymple, *The Fourth Dimension and Non-Euclidean Geometry in Modern Art* (Cambridge, MA: MIT Press, 2013).

Hoffmann, Banesh, with Helen Dukas, *Albert Einstein: Creator and Rebel* (New York: Viking, 1972).

Holton, Gerald, and Yehuda Elkana, editors, *Albert Einstein: Historical and Cultural Perspectives* (Princeton, NJ: Princeton University Press, 1982).

Howard, Don, "Albert Einstein as a Philosopher of Science," *Physics Today* 58 (2005): 34–40.

*———, "Einstein on Locality and Separability," *Studies in History and Philosophy of Science* 16 (1987): 171–201.

*———, "Who Invented the Copenhagen Interpretation? A Study in Mythology," *Philosophy of Science* 71 (2004): 669–682.

Howard, Don, and John Stachel, editors, *Einstein: The Formative Years 1879–1909* (Boston: Birkhäuser, 2000).

Isaacson, Walter, *Einstein: His Life and Universe* (New York: Simon and Schuster, 2008).

Jammer, Max, *The Conceptual Development of Quantum Mechanics* (New York: McGraw-Hill, 1966).

Kaku, Michio, *Einstein's Cosmos: How Albert Einstein's Vision Transformed Our Understanding of Space and Time* (New York: W. W. Norton, 2005).

Kragh, Helge, *Quantum Generations: A History of Physics in the Twentieth Century* (Princeton: Princeton University Press, 1999).

Mach, Ernst, *The Science of Mechanics: A Critical and Historical Exposition of Its Principles*, translated by Thomas McCormack (Chicago: Open Court, 1897).

———, *Space and Geometry*, translated by Thomas McCormack (Chicago: Open Court, 1897).

Mehra, Jagesh, *Erwin Schrödinger and the Rise of Wave Mechanics, Part 1: Schrödinger in Vienna and Zurich, 1887–1925*, The Historical Development of Quantum Theory, volume 5 (New York: Springer, 1987).

Moore, Walter, *Schrödinger: Life and Thought* (New York: Cambridge University Press, 1982).

Pais, Abraham, *Subtle Is the Lord . . . : The Science and the Life of Albert Einstein* (Oxford: Oxford University Press, 1982).

Parker, Barry, *Einstein's Dream: The Search for a Unified Theory of the Universe* (New York: Plenum, 1986).

———, *Search for a Supertheory: From Atoms to Superstrings* (New York, Plenum, 1987).

*Pesic, Peter, *Beyond Geometry: Classic Papers from Riemann to Einstein* (New York: Dover, 2006).

Pickover, Clifford, *Surfing Through Hyperspace: Understanding Higher Universes in Six Easy Lessons* (New York: Oxford University Press, 1999).

*Putnam, Hilary, "A Philosopher Looks at Quantum Mechanics (Again)," *British Journal for the Philosophy of Science* 26 (2005): 615–634.

Sayen, Jamie, *Einstein in America* (New York: Crown, 1985).

*Schrödinger, Erwin, *Space-Time Structure* (Cambridge: Cambridge University Press, 1950).

———, *My View of the World*, translated by Cecily Hastings (Woodbridge, CT: Ox Bow Press, 1983).

———, *What Is Life?* (Cambridge: Cambridge University Press, 1950).

Seelig, Carl, *Albert Einstein: A Documentary Biography*, translated by Mervyn Savill (London: Staples Press, 1956).

Smith, Peter D., *Einstein: Life and Times* (London: Haus Publishing, 2005).

Stachel, John, *Einstein from "B" to "Z"* (Boston: Birkhäuser, 2002).

———, "History of Relativity," in *Twentieth Century Physics*, vol. 1, edited by Laurie Brown et al. (New York: American Institute of Physics Press, 1995).

Thirring, Walter, *Cosmic Impressions: Traces of God in the Laws of Nature*, translated by Margaret A. Schellenberg (Philadelphia: Templeton Foundation Press, 2007).

*Vizgin, Vladimir, "The Geometrical Unified Field Theory Program," in *Einstein and the History of General Relativity*, edited by Don Howard and John Stachel, 300–314 (Boston: Birkhäuser, 1989).

*———, *Unified Field Theories: In the First Third of the 20th Century*, translated by J. B. Barbour (Boston: Birkhäuser, 1994).

Weinberg, Steven, *Dreams of a Final Theory: The Scientist's Search for the Ultimate Laws of Nature* (New York: Vintage, 1992).

Weyl, Hermann, *Space, Time, Matter* (New York: Dover, 1950).

Notes

INTRODUCTION: ALLIES AND ADVERSARIES

1. Erwin Schrödinger, "The New Field Theory," January 1947, Albert Einstein Duplicate Archive, Princeton, NJ, 22-152.

2. "Unifying the Cosmos," *New York Times*, February 16, 1947.

3. Elihu Lubkin, "Schrödinger's Cat," *International Journal of Theoretical Physics* 18, no. 8 (1979): 520.

4. Hilary Putnam, personal correspondence with the author, August 4, 2013.

5. Walter Thirring, *Cosmic Impressions: Traces of God in the Laws of Nature*, trans. Margaret A. Schellenberg (Philadelphia: Templeton Foundation Press, 2007), 54.

6. Ibid., 55.

7. "Einstein Tribute to Schroedinger," *Irish Times*, June 29, 1943, 3.

8. Albert Einstein, "Statement to the Press," February 1947, Albert Einstein Duplicate Archive, 22-146.

9. Albert Einstein, quoted in "Einstein's Comment on Schroedinger Claim," *Irish Press*, February 27, 1.

10. Myles na gCopaleen [Brian O' Nolan], "Cruiskeen Lawn," *Irish Times*, March 10, 1947, 4.

11. John Moffat, *Einstein Wrote Back: My Life in Physics* (Toronto: Thomas Allen, 2010), 67.

12. Peter Freund, *A Passion for Discovery* (Hackensack, NJ: World Scientific, 2007), 5–6.

CHAPTER ONE: THE CLOCKWORK UNIVERSE

1. Albert Einstein, *Autobiographical Notes*, trans. and ed. Paul Arthur Schilpp (La Salle, IL: Open Court, 1979), 9.

2. John Casey, *The First Six Books of the Elements of Euclid* (Dublin: Hodges, Figgis, 1885), 6.

3. Einstein's ideas were anticipated by British mathematician William Kingdon Clifford, who in 1870 made use of Riemann's description of curvature to try to model matter through geometry. Clifford also translated Riemann's treatise into English, publishing his rendition in 1873. However, it wasn't until well after Einstein developed general relativity in 1915

that Clifford's contributions to the study of how matter and geometry were connected would become widely recognized.

4. Ernst Mach, "Die Leitgedanken meiner naturwissenschaftlichen Errkenntnislehre und ihre Aufnahme durch die Zeitgenossen," *Scientia* 8 (1910), trans. as "The Guiding Principles of My Scientific Theory of Knowledge and Its Reception by My Contemporaries," in S. Toulmin, ed., *Physical Reality* (New York: Harper & Row, 1970), 37–38.

5. Erwin Schrödinger, Antrittsrede des Herrn Schrödinger, *Sitz. Ber. Preuss. Akad. Wiss.* (Berlin) 1929, p. C, quoted in Jagesh Mehra and Helmut Rechenberg, *Erwin Schrödinger and the Rise of Wave Mechanics, Part 1: Schrödinger in Vienna and Zurich, 1887–1925,* The Historical Development of Quantum Theory, volume 5 (New York: Springer, 1987), 81.

6. The reasons for Hasenöhrl's near-miss are discussed in Stephen Boughn, "Fritz Hasenöhrl and $E = mc^2$," *European Physical Journal H* 38 (2013): 261–278.

7. Interview with Dr. Hans Thirring by Thomas S. Kuhn in Vienna, Austria, April 4, 1963, Archive for the History of Quantum Physics, American Philosophical Society, Philadelphia, PA.

8. Einstein, *Autobiographical Notes,* 15.

9. Albert Einstein to Anna Keller Grossmann, reprinted in Carl Seelig, *Albert Einstein: A Documentary Biography,* trans. Mervyn Savill (London: Staples Press, 1956), 208.

10. Max Talmey, "Einstein as a Boy Recalled by a Friend," *New York Times,* February 10, 1929, 11.

11. Max von Laue, quoted in Seelig, *Albert Einstein,* 78.

12. Hermann Minkowski, address delivered at the Eightieth Assembly of German Natural Scientists and Physicians, September 21, 1908.

CHAPTER TWO: THE CRUCIBLE OF GRAVITY

1. *Punch,* November 19, 1919, 422, cited in Alistair Sponsel, "Constructing a 'Revolution in Science': The Campaign to Promote a Favourable Reception for the 1919 Solar Eclipse Experiments," *British Journal for the History of Science* 35, no. 4 (2002): 439.

2. Jagdish Mehra and Helmut Rechenberg, *Erwin Schrödinger and the Rise of Wave Mechanics, Part 1: Schrödinger in Vienna and Zurich, 1887–1925,* The Historical Development of Quantum Theory, volume 5 (New York: Springer, 1987), 166.

3. George de Hevesy to Ernest Rutherford, October 14, 1913. Rutherford Papers, University of Cambridge, quoted in Ronald W. Clark, *Einstein: The Life and Times* (New York: World Publishing, 1971), 158.

4. Erwin Schrödinger, *Space-Time Structure* (Cambridge: Cambridge University Press, 1963), 1.

5. Albert Einstein, speech given in Kyoto, Japan, on December 14, 1922, quoted in Engelbert L. Schücking and Eugene J. Surowitz, "Einstein's Apple," unpublished manuscript, 2013.

6. Albert Einstein to Arnold Sommerfeld, October 29, 1912, in Albert Einstein, *The Collected Papers of Albert Einstein,* vol. 5, *The Swiss Years: Correspondence, 1902–1914,* English translation supplement, ed. Don Howard, trans. Anna Beck (Princeton, NJ: Princeton University Press, 1995), Doc. 421.

7. Carl Seelig, *Albert Einstein: A Documentary Biography,* trans. Mervyn Savill (London: Staples Press, 1956), 108.

8. Albert Einstein to Paul Ehrenfest, January 1916, in Seelig, *Albert Einstein,* 156.

9. Richard Feynman, *"Surely You're Joking, Mr. Feynman!": Adventures of a Curious Character* (New York: Norton, 2010), 58.

10. Walter Moore, *Schrödinger: Life and Thought* (New York: Cambridge University Press, 1982), 105.

11. Erwin Schrödinger, translated and quoted in Alex Harvey, "How Einstein Discovered Dark Energy," 2012, http://arxiv.org/abs/1211.6338.

12. Albert Einstein, "Bemerkung zu Herrn Schrödingers Notiz Über ein Lösungssystem der allgemein kovarianten Gravitationsgleichungen," *Physikalische Zeitschrift* 19 (1918): 165–166, translated and edited by M. Janssen et al. in *The Collected Papers of Albert Einstein,* vol. 7, *The Berlin Years: Writings, 1918–1921* (Princeton: Princeton University Press, 2002), doc. 3.

13. Harvey, "How Einstein Discovered Dark Energy."

14. Ben Almassi, "Trust in Expert Testimony: Eddington's 1919 Eclipse Expedition and the British Response to General Relativity," *Studies in History and Philosophy of Science Part B* 40, no. 1 (2009): 57–67.

15. Ibid.

16. "Eclipse Showed Gravity Variation," *New York Times,* November 8, 1919, 6.

17. Ibid.

18. "Revolution in Science . . . New Theory of the Universe . . . Newtonian Ideas Overthrown," *Times* (London), November 7, 1919, 1.

19. Erwin Schrödinger, *Space-Time Structure* (Cambridge: Cambridge University Press, 1963), 2.

20. Albert Einstein, "On the Method of Theoretical Physics" (1933 lecture at Oxford), translated by S. Bargmann in *Albert Einstein: Ideas and Opinions* (New York: Bonanza Books, 1954), 270–276.

21. David Hilbert, MacTutor online biography, University of St. Andrews, http://www-history.mcs.st-andrews.ac.uk/Biographies/Hilbert.html.

22. Albert Einstein to Hermann Weyl, March 8, 1918, in Albert Einstein, *The Collected Papers of Albert Einstein*, vol. 8, *The Berlin Years: Correspondence, 1914–1918*, English translation supplement, ed. Klaus Hentschel, trans. Ann M. Hentschel (Princeton, NJ: Princeton University Press, 1998).

23. Daniela Wünsch, *Der Erfinder der 5. Dimension, Theodor Kaluza* (Göttingen: Termessos, 2007), 66.

24. Theodor Kaluza Jr., interviewed in *NOVA: What Einstein Never Knew*, PBS, originally broadcast October 22, 1985.

25. Arthur S. Eddington, "A Generalisation of Weyl's theory of the Electromagnetic and Gravitational Fields," *Proceedings of the Royal Society of London, Ser. A* 99 (1921): 104–122.

CHAPTER THREE: MATTER WAVES AND QUANTUM JUMPS

1. Omar Khayyam, *The Rubaiyat of Omar Khayyam*, trans. Edward Fitzgerald (New York: Dover, 2011).

2. Jagdish Mehra and Helmut Rechenberg, *Erwin Schrödinger and the Rise of Wave Mechanics, Part 1: Schrödinger in Vienna and Zurich, 1887–1925*, The Historical Development of Quantum Theory, volume 5 (New York: Springer, 1987), 408.

3. Erwin Schrödinger, *My View of the World*, trans. Cecily Hastings (Woodbridge, CT: Ox Bow Press, 1983), 7.

4. Baruch Spinoza, *Ethics*, in *The Collected Writings of Spinoza*, vol. 1, trans. Edwin Curley (Princeton: Princeton University Press, 1985).

5. Albert Einstein, quoted in "Einstein Believes in 'Spinoza's God,'" *New York Times*, April 25, 1929, 1.

6. Albert Einstein, "Religion and Science," *New York Times Magazine*, November 9, 1930, SM1.

7. Schrödinger, *My View of the World*, 21.

8. W. Heitler, "Erwin Schrödinger Obituary" *Roy. Soc. Obit.* 7 (1961): 223–234.

9. Wolfgang Pauli, quoted in Werner Heisenberg, *Physics and Beyond* (New York: Harper and Row, 1971), 25–26.

10. Peter Freund, *A Passion for Discovery* (Hackensack, NJ: World Scientific, 2007), 162.

11. Erwin Schrödinger to Albert Einstein, November 3, 1925, Albert Einstein Duplicate Archive, Princeton, NJ, 22-004.

12. Schrödinger, *My View of the World*, p. 54.

13. Hermann Weyl, reported by Abraham Pais, *Inward Bound: Of Matter and Forces in the Physical World* (New York: Oxford University Press, 1988), 252.

14. Arnold Sommerfeld to Erwin Schrödinger, February 3, 1926, reported in Mehra and Rechenberg, *Erwin Schrödinger and the Rise of Wave Mechanics*, 537.

15. Interview with Annemarie Schrödinger by Thomas S. Kuhn in Vienna, Austria, April 5, 1963, Archive for the History of Quantum Physics, American Philosophical Society, Philadelphia, PA.

16. Erwin Schrödinger to Albert Einstein, April 23, 1926, Albert Einstein Duplicate Archive, 22-014.

17. Erwin Schrödinger to Niels Bohr, May 24, 1924, quoted and translated in O. Darrigol, "Schrödinger's Statistical Physics and Some Related Themes," in M. Bitbol and O. Darrigol, eds., *Erwin Schrödinger, Philosophy and the Birth of Quantum Mechanics* (Gif-sur-Yvette, France: Editions Frontières, 1992).

18. Albert Einstein to Max Born, December 4, 1926, in *Albert Einstein-Max Born, Briefwechsel (Correspondence)*, ed. Max Born (Munich, 1969), 129, quoted in Alice Calaprice and Trevor Lipscombe, *Albert Einstein: A Biography* (Westport, CT: Greenwood Press, 2005), 92.

19. Albert Einstein to Max Born, May 1927, reprinted in A. Einstein, H. Born, and M. Born, *Albert Einstein, Hedwig und Max Born, Briefwechsel: 1916–1955 / kommentiert von Max Born; Geleitwort von Bertrand Russell; Vorwort von Werner Heisenberg* (Frankfurt am Main: Edition Erbrich, 1982), 136, quoted and translated in Hubert Goenner, "On the History of Unified Field Theories," *Living Reviews in Relativity*, 2004, http://relativity.livingreviews.org/Articles/lrr-2004-2/download/lrr-2004-2Color.pdf.

20. Albert Einstein to Erwin Schrödinger, May 31, 1928, Albert Einstein Duplicate Archive, 22-022, quoted and translated in G. G. Emch, *Mathematical and Conceptual Foundations of 20th-Century Physics* (Amsterdam: North Holland, 2000), 295.

21. Abraham Pais, *Einstein Lived Here* (New York: Oxford University Press, 1994), 43.

CHAPTER FOUR: THE QUEST FOR UNIFICATION

1. Interview with Annemarie Schrödinger by Thomas S. Kuhn in Vienna, Austria, April 5, 1963, Archive for the History of Quantum Physics, American Philosophical Society, Philadelphia, PA.

2. Paul Heyl, "What Is an Atom?," *Scientific American* 139 (July 1928): 9–12.

3. "Current Magazines," *New York Times*, July 1, 1928.

4. Albert Einstein, quoted in "Einstein Declares Women Rule Here," *New York Times*, July 8, 1921.

5. "Woman Threatens Prof. Einstein's Life," *New York Times*, February 1, 1925.

6. "A Deluded Woman Threatens Krassin and Professor Einstein," *The Age* (Melbourne, Australia), February 3, 1925, 9.

7. Wythe Williams, "Einstein Distracted by Public Curiosity," *New York Times*, February 4, 1929.

8. Einstein to Zangger, end of May 1928, Einstein Archives, Hebrew University of Jerusalem, call no. 40-069, translated and quoted in Tilman Sauer, "Field Equations in Teleparallel Spacetime: Einstein's *Fernparallelismus* Approach Towards Unified Field Theory," *Historia Mathematica* 33 (2006): 404–405.

9. "Einstein Extends Relativity Theory," *New York Times*, January 12, 1929, 1.

10. Albert Einstein, quoted in "Einstein Is Amazed at Stir over Theory; Holds 100 Journalists at Bay for a Week," *New York Times*, January 19, 1929.

11. Albert Einstein, quoted in "News and Views," *Nature*, February 2, 1929, reprinted in Hubert Goenner, "On the History of Unified Field Theories," in *Proceedings of the Sir Arthur Eddington Centenary Symposium*, edited by V. de Sabbata and T. M. Karade, 1:176–196 (Singapore: World Scientific, 1984).

12. H. H. Sheldon, quoted in "Einstein Reduces All Physics to 1 Law," *New York Times*, January 25, 1929.

13. "Einstein Is Viewed as Near the Mystic," *New York Times*, February 4, 1929.

14. Will Rogers, "Will Rogers Takes a Look at the Einstein Theory," *New York Times*, February 1, 1929.

15. "Byproducts: Some Parallel Vectors," *New York Times*, February 3, 1929.

16. Wolfgang Pauli, "[Besprechung von] Band 10 der Ergebnisse der exakten Naturwissenschaften," *Ergebnisse der exakten Naturwissenschaften* 11 (1931): 186, quoted and translated in Goenner, "On the History of Unified Field Theories."

17. "Einstein Flees Berlin to Avoid Being Feted," *New York Times*, March 13, 1929.

18. "Einstein Is Found Hiding on Birthday," *New York Times*, March 14, 1929.

19. Walter Moore, *Schrödinger: Life and Thought* (New York: Cambridge University Press, 1982), 242.

20. Paul Dirac, quoted in "Erwin Schrödinger," Archive for the History of Quantum Physics.

21. Albert Einstein, quoted in "Einstein Affirms Belief in Causality," *New York Times*, March 16, 1931, 1.

22. "Physicists Scan Cause to Effect with Skepticism," *Christian Science Monitor*, November 13, 1931, 8.

23. Albrecht Fölsing, *Albert Einstein: A Biography*, trans. Ewald Osers (New York: Penguin, 1997), 617.

24. Moore, *Schrödinger*, 255.

CHAPTER FIVE: SPOOKY CONNECTIONS AND ZOMBIE CATS

1. Annemarie Schrödinger, reported in Walter Moore, *Schrödinger: Life and Thought* (New York: Cambridge University Press, 1982), 265.

2. "Relative Tide and Sand Bars Trap Einstein; He Runs His Sailboat Aground at Old Lyme," *New York Times*, August 4, 1935, 1.

3. Don Duso, reported in Sandi Fairbanks, *All Points North Magazine*, Summer 2008, www.apnmag.com/summer_2008/fairbanks_einstein.php.

4. Albert Einstein to Elisabeth, Queen of Belgium, autumn 1935, quoted in Ronald Clark, *Einstein: The Life and Times* (New York: World Publishing, 1971), 529.

5. Albert Einstein, quoted in "Einsteinhaus in Caputh," www.einstein-sommerhaus.de.

6. Moore, *Schrödinger*, 294.

7. Erwin Schrödinger to Albert Einstein, June 7, 1935, quoted and translated in Don Howard, "Revisiting the Einstein-Bohr Dialogue," *Iyyun: The Jerusalem Philosophical Quarterly* 56 (January 2007): 21–22.

8. Albert Einstein to Erwin Schrödinger, June 19, 1935, Albert Einstein Duplicate Archive, Princeton, NJ, 22-047.

9. Ibid.

10. "Einstein Attacks Quantum Theory," *New York Times*, May 4, 1935.

11. Albert Einstein to Erwin Schrödinger, August 8, 1935, Albert Einstein Duplicate Archive, 22-049.

12. Ibid.

13. Erwin Schrödinger to Albert Einstein, August 19, 1935, Albert Einstein Duplicate Archive, 22-051.

14. Ruth Braunizer, reported by Leonhard Braunizer, personal correspondence with the author, May 6, 2014.

15. Albert Einstein to Erwin Schrödinger, September 4, 1935, Albert Einstein Duplicate Archive, 22-052.

16. Erwin Schrödinger, "Die Gegenwärtigen Situation in der Quantenmechanik," *Die Naturwissenschaften* 23 (1935): 807–812, 824–828, quoted and translated in Arthur Fine, *The Shaky Game: Einstein, Realism and the Quantum Theory* (Chicago: University of Chicago Press, 1986), 65.

17. Erwin Schrödinger, "Indeterminism and Free Will," *Nature*, July 4, 1936.

18. Ibid.

19. Interview with Annemarie Schrödinger by Thomas S. Kuhn in Vienna, Austria, April 5, 1963, Archive for the History of Quantum Physics, American Philosophical Society, Philadelphia, PA.

20. Helge Kragh, *Quantum Generations: A History of Physics in the Twentieth Century* (Princeton: Princeton University Press, 1999), 218–229.

21. Jamie Sayen, *Einstein in America* (New York: Crown, 1985), 147.

22. Lucien Aigner, "A Book May Be Written, a Shoe Made—But a Theory—It's Never Finished," *Christian Science Monitor*, December 14, 1940, 3.

23. Nathan Rosen, "Reminiscences," in Gerald Holton and Yehuda Elkana, eds., *Albert Einstein: Historical and Cultural Perspectives* (Princeton: Princeton University Press, 1982), 406.

24. Erwin Schrödinger, "Confession to the Führer," *Graz Tagespost*, March 30, 1938, quoted and translated in Moore, *Schrödinger: Life and Thought*, 337.

25. Erwin Schrödinger, quoted in "History of the Dublin Institute for Advanced Studies: 1935–1940: Formation of the School," Dublin Institute for AdvancedStudies,www.dias.ie/index.php?option=com_content&view=article &id=804:theoreticalhistory1935-1940.

26. Erwin Schrödinger, unpublished manuscript, Dublin Institute for Advanced Studies Archive, quoted in Moore, *Schrödinger: Life and Thought*, 348.

27. Brian Fallon, *An Age of Innocence: Irish Culture, 1930–1960* (London: Palgrave Macmillan, 1998), 14.

CHAPTER SIX: LUCK OF THE IRISH

1. Walter Thirring, *Cosmic Impressions: Traces of God in the Laws of Nature*, translated by Margaret A. Schellenberg (Philadelphia: Templeton Foundation Press, 2007), 55.

2. Nicola Tallant, "Dev Tricked Public into Investing in Irish Press, File Reveals," *Irish Independent*, October 31, 2004, 1.

3. L. Mac G., "A Professor at Home," *Irish Press*, November 1, 1940, 5.

4. "People and Places," *Irish Press*, August 11, 1942, 2.

5. "The 'Atom Man' at Home: Dr. Erwin Schrödinger Takes a Day Off," *Irish Press*, February 1, 1946, 7.

6. Gespräch mit Ruth Braunizer über Erwin Schrödinger (interview with Ruth Braunizer about Erwin Schrödinger), Österreichische Mediathek, 1997, http://www.oesterreich-am-wort.at/treffer/atom/14957620 -36E-00084-00000AF8-1494EDB5.

7. Ruth Braunizer, "Memories of Dublin—Excerpts from Erwin Schrödinger's Diaries," in Gisela Holfter, ed., *German-Speaking Exiles in Ireland 1933–1945* (Amsterdam: Rodopi, 2006), 265.

8. Albert Einstein, quoted in Robert P. Crease, *The Great Equations: Breakthroughs in Science from Pythagoras to Heisenberg* (New York: W. W. Norton, 2010), 197.

9. Leopold Infeld, "Visit to Dublin," *Scientific American* 181, no. 4 (October 1949): 11.

10. Erwin Schrödinger, "Some Thoughts on Causality," *Irish Times*, November 15, 1939, 5.

11. Myles na gCopaleen [Brian O'Nolan], reported in Paddy Leahy, "How Myles na gCopaleen Belled Schrödinger's Cat," *Irish Times*, February 22, 2001, 15.

12. Myles na gCopaleen [Brian O'Nolan], "Cruiskeen Lawn," *Irish Times*, August 3, 1942, 3.

13. Flann O'Brien [Brian O'Nolan], *The Third Policeman* (Chicago: Dalkey Archive Press, 2006), 116.

14. "Observer Says," *Irish Press*, November 9, 1943, 3.

15. Ibid.

16. "Famous Physicist's Memory to Be Honoured by Special Stamp," *Irish Press*, November 6, 1943, 1.

17. Albert Einstein to Hans Muehsam, early summer 1942, quoted in Carl Seelig, *Albert Einstein: A Documentary Biography*, translated by Mervyn Savill (London: Staples Press, 1956), 230.

18. Peter Seyyfart, "Einstein, Mann Popular at Princeton; Students 'Praise' Them in Jingles," *Milwaukee Journal*, August 12, 1939.

19. Léon Rosenfeld to Friedrich Herneck, 1962, published in F. Herneck, *Einstein und sein Weltbild* (Berlin: Buchverlag der Morgen, 1976), 280.

20. Albert Einstein, address to the American Scientific Congress, May 15, 1940, reported in William L. Laurence, "Einstein Baffled by Cosmos Riddle," *New York Times*, May 16, 1940, 23.

21. Institute for Advanced Study School of Mathematics, Confidential Memo, April 19, 1945, IAS Archive, Princeton, NJ.

22. Albert Einstein and Wolfgang Pauli, "On the Non-Existence of Regular Stationary Solutions of Relativistic Field Equations," *Annals of Mathematics* 44 (April 1943): 13.

23. Michael Lawlor, "Forward from Einstein," *Irish Press*, February 1, 1943, 2.

24. "Scholars Acclaim His Theory," *Irish Press*, February 2, 1943, 2.

25. "Science: Schroedinger," *Time*, April 5, 1943.

26. "Einstein's Comment on Schroedinger Theory," *Irish Press*, April 10, 1943, 1.

27. "Einstein Tribute to Schroedinger," *Irish Times*, June 29, 1943, 3.

28. George Prior Woollard, "Transcontinental Gravitational and Magnetic Profile of North America and Its Relation to Geologic

Structure," *Geological Society of America Bulletin* 54, no. 6 (June 1, 1943): 747–789.

29. "Schroedinger's New Theory Confirmed," *Irish Press,* June 28, 1943, 1.

30. Erwin Schrödinger to Albert Einstein, August 13, 1943, Albert Einstein Duplicate Archive, 22-075.

31. Albert Einstein to Erwin Schrödinger, September 10, 1943, Albert Einstein Duplicate Archive, 22-076.

32. Erwin Schrödinger to Albert Einstein, October 31, 1943, Albert Einstein Duplicate Archive, 22-088.

33. Albert Einstein to Erwin Schrödinger, December 14, 1943, Albert Einstein Duplicate Archive, 22-090.

34. Reported in Walter Moore, *Schrödinger: Life and Thought* (New York: Cambridge University Press, 1982), 418. Moore has speculated that because Schrödinger always wanted a son, he hoped that she would get pregnant on the chance she would have a boy.

35. John Gribbin, *Erwin Schrödinger and the Quantum Revolution* (Hoboken, NJ: Wiley, 2013), 285.

36. Matthew Benjamin, "Catcher, Spy: Moe Berg," *US News and World Report,* January 27, 2003.

CHAPTER SEVEN: PHYSICS BY PUBLIC RELATIONS

1. Walter Winchell, "Scientists See Steel Block Melted by Light Beam," *Spartanburg Herald Journal,* May 23, 1948, A4.

2. D. M. Ladd, Office Memorandum to the Director, Federal Bureau of Investigation, February 15, 1950, *FBI Records: The Vault*, http://vault.fbi.gov /Albert Einstein.

3. Robin Pogrebin, "Love Letters by Einstein at Auction." *New York Times,* June 1, 1998.

4. Reported in Carl Seelig, *Albert Einstein: A Documentary Biography,* trans. Mervyn Savill (London: Staples Press, 1956), 115.

5. "A Summary of Fianna Fáil's Self Claimed Achievements as Used by the Party During the General Election of 1948," University College Dublin Archive P150/2756, reprinted in Diarmaid Ferriter, *Judging Dev: A Reassessment of the Life and Legacy of Éamon de Valera* (Dublin: Royal Irish Academy Press, 2007), 296.

6. James Dillon, "Constituent School of the Dublin Institute for Advanced Studies—Motion," *Dáil Éireann Proceedings* 104 (February 13, 1947).

7. Wolfgang Pauli, quoted in Vladimir Vizgin, *Unified Field Theories: In the First Third of the 20th Century,* trans. J. B. Barbour (Boston: Birkhäuser, 1994), 218.

8. Albert Einstein to Erwin Schrödinger, January 22, 1946, Albert Einstein Duplicate Archive, Princeton, NJ, 22-093.

9. Erwin Schrödinger to Albert Einstein, February 19, 1946, Albert Einstein Duplicate Archive, 22-094.

10. Erwin Schrödinger to Albert Einstein, March 24, 1946, Albert Einstein Duplicate Archive, 22-102.

11. Albert Einstein to Erwin Schrödinger, April 7, 1946, Albert Einstein Duplicate Archive, 22-103.

12. Erwin Schrödinger to Albert Einstein, June 13, 1946, Albert Einstein Duplicate Archive, 22-107.

13. Albert Einstein to Erwin Schrödinger, July 16, 1946, Albert Einstein Duplicate Archive, 22-109.

14. Albert Einstein to Erwin Schrödinger, January 27, 1947, Albert Einstein Duplicate Archive, 22-136.

15. William Rowan Hamilton, quoted in Robert Percival Graves, *Life of Sir William Rowan Hamilton* (Dublin: Hodges, Figgis, 1882).

16. Erwin Schrödinger, "The Final Affine Field-Laws," Address to the Royal Irish Academy, January 27, 1947, Albert Einstein Duplicate Archive, 22-143.

17. Ibid.

18. Erwin Schrödinger, quoted in "Dr. Schroedinger: Einstein Theory of Relativity," *Irish Press*, January 28, 1947, 5.

19. "Dublin Man Outdoes Einstein," *Christian Science Monitor*, January 31, 1947, 13.

20. Erwin Schrödinger to Albert Einstein, February 3, 1947, Albert Einstein Duplicate Archive, 22-138.

21. "Science: Einstein Stopped Here," *Time*, February 10, 1947.

22. John L. Synge, "Letter to the Editor," *Time*, March 3, 1947.

23. Petros S. Florides, "John Lighton Synge," *Biographical Memoirs of Fellows of the Royal Society* 54 (December 2008): 401.

24. Nichevo [R. M. Smyllie], "Higher Maths," *Irish Times*, March 22, 1947, 7.

25. S. McC., "And Now Cosmic Physics," *Tuam Herald*, April 12, 1947.

26. William L. Laurence to Albert Einstein, February 7, 1947, Albert Einstein Duplicate Archive, 22-141.

27. "Einstein Declines Comment," *New York Times*, January 30, 1947.

28. "Einstein's Theory Reportedly Widened," *New York Times*, January 30, 1947.

29. "Unifying the Cosmos," *New York Times*, February 16, 1947.

30. Jacob Landau to Albert Einstein, February 18, 1947, Albert Einstein Duplicate Archive, 22-149.

31. Albert Einstein, "Statement to the Press," February 1947, Albert Einstein Duplicate Archive, 22-146.

32. Erwin Schrödinger, quoted in "Schroedinger Replies to Einstein," *Irish Press*, March 1, 1947, 7.

33. Peter Freund, *A Passion for Discovery* (Hackensack, NJ: World Scientific, 2007), 5.

34. Myles na gCopaleen [Brian O'Nolan], "Cruiskeen Lawn," *Irish Times*, March 10, 1947, 4.

35. John Archibald Wheeler, interview with the author, Princeton, November 5, 2002.

36. "Einstein Leaves Hospital," *New York Times*, January 14, 1949.

37. William L. Laurence, "World Scientists Hail Einstein at 70," *New York Times*, March 13, 1949.

CHAPTER EIGHT: THE LAST WALTZ: EINSTEIN'S AND SCHRÖDINGER'S FINAL YEARS

1. Lincoln Barnett, "U.S. Science Holds Its Biggest Powwow and Finds It Has a New Einstein Theory to Ponder—The Meaning of Einstein's New Theory," *Life*, January 9, 1950.

2. Datus Smith to Lincoln Barnett, January 6, 1950, Princeton University Press Archive, Box 7, Princeton University Library; Lincoln Barnett to Datus Smith, January 18, 1950, Princeton University Press Archive.

3. Datus Smith to Lincoln Barnett, January 23, 1950, Princeton University Press Archive.

4. Frances Hagemann to Albert Einstein (copy to Herbert Bailey), January 14, 1950, Princeton University Press Archive.

5. Herbert Bailey to Frances Hagemann, January 18, 1950, Princeton University Press Archive.

6. Frances Hagemann to Herbert Bailey (copy to Albert Einstein), January 26, 1950, Princeton University Press Archive.

7. *Irish Times,* January 2, 1950, 5.

8. William L. Laurence, "Einstein Publishes His 'Master Theory,'" *New York Times*, February 15, 1950.

9. Robert Oppenheimer, "On Albert Einstein," *New York Review of Books*, March 17, 1966.

10. Erwin Schrödinger, Interviewed in "Einstein Has New Theory of Laws of Gravitation," *Irish Press*, December 26, 1949, 1.

11. Erwin Schrödinger, *Space-Time Structure* (Cambridge: Cambridge University Press, 1963), 114.

12. Ibid., 116.

13. Albert Einstein to Erwin Schrödinger, September 3, 1950, Albert Einstein Duplicate Archive, 22-171.

14. Erwin Schrödinger to Albert Einstein, May 15, 1953, Albert Einstein Duplicate Archive, 22-210.

15. Albert Einstein to Erwin Schrödinger, June 9, 1953, Albert Einstein Duplicate Archive, 22-212.

16. Robert Romer, "My Half Hour with Einstein," *Physics Teacher* 43 (2005): 35.

17. Albert Einstein, quoted in Werner Heisenberg, *Encounters with Einstein* (Princeton, NJ: Princeton University Press, 1989), 121.

18. Eugene Shikhovtsev, "Biographical Sketch of Hugh Everett, III," edited by Kenneth W. Ford, http://space.mit.edu/home/tegmark/everett/everett.html.

19. Arthur I. Miller, *Deciphering the Cosmic Number: The Strange Friendship of Wolfgang Pauli and Carl Jung* (New York: Norton, 2010), 269.

20. Wolfgang Pauli to George Gamow, March 1, 1958, reported in Miller, *Deciphering the Cosmic Number,* 263.

21. Erwin Schrödinger, 1942 poem, translated by Arnulf Braunizer, reprinted in Amir Aczel, *Present at the Creation: Discovering the Higgs Boson* (New York: Random House, 2010), 33.

22. Ursula K. Le Guin, interviewed by Irv Broughton, *Conversations with Ursula K. Le Guin* (Jackson: University Press of Mississippi, 2008), 59.

23. Roland Orzabal, personal correspondence with the author, September 17, 2013.

24. Klaus Taschwer, "Der Streit um Schrödingers Kiste," *Der Standard* (Austria), December 19, 2007.

25. "Schrödingers Erbe: Gerichtlicher Streit beigelegt," Österreichischen Rundfunk, May 13, 2009.

CONCLUSION: BEYOND EINSTEIN AND SCHRÖDINGER:
THE ONGOING SEARCH FOR UNITY

1. "Laid-Back Surfer Dude May Be Next Einstein," FoxNews.com, November 16, 2007, www.foxnews.com/story/2007/11/16/laid-back-surfer-dude-may-be-next-einstein.

2. "Autistic Boy, 12, with Higher IQ than Einstein Develops His Own Theory of Relativity," *Daily Mail Online,* March 24, 2011, www.dailymail.co.uk/news/article-1369595/Jacob-Barnett-12-higher-IQ-Einstein-develops-theory-relativity.html.

3. "Will the Next Einstein Be a Computer?," KitGuru Online Forum, www.kitguru.net/channel/science/jules/will-the-next-einstein-be-a-computer.

4. Kane Fulton, "Ubuntu on Android May Help Find Next Einstein," TechRadar, June 18, 2013, wwwtechradar.com/us/news/software/operating-systems/-ubuntu-on-android-may-help-find-next-einstein--1159142.

5. Tamar Lewin, "No Einstein in Your Crib? Get a Refund!," *New York Times,* October 24, 2009, A1.

6. Ian Sample, "Faster Than Light Particles Found, Claim Scientists," *The Guardian,* September 22, 2011.

7. Antonio Ereditato, press release, OPERA experiment, September 23, 2011.

8. "Neutrino Jokes Hit Twittersphere Faster Than the Speed of Light," *Los Angeles Times,* September 24, 2011.

9. Corrigan Brothers and Pete Creighton, "Neutrino Song," October 10, 2011, www.youtube.com/watch?v=vpMY84T8WY0.

10. Sergio Bertolucci, press release, CERN, June 8, 2012.

Index

Courtesy of the University of the Sciences

PAUL HALPERN is a professor of physics at the University of the Sciences in Philadelphia and the author of fourteen popular science books. He lives near Philadelphia, Pennsylvania.